Ben Stacy Jerrik (Ed.)

(152576) 1994 NR2

Ben Stacy Jerrik (Ed.)

(152576) 1994 NR2

Minor Planet, Asteroid Belt, Asteroid, , Trojan (Astronomy)

Part Press

Imprint

Publisher:
Part Press is a trademark of
International Book Market Service Ltd., 17 Rue Meldrum, Beau Bassin, 1713-01 Mauritius
Email: info@bookmarketservice.com
Website: www.bookmarketservice.com

Published in 2013

Printed in: U.S.A., U.K., Germany. This book was not produced in Mauritius.

ISBN: 978-613-8-72910-5

Contents

Articles

References

(152576) 1994 NR2

(152576) 1994 NR_2 is a main-belt minor planet. It was discovered by Henri Debehogne and Eric Walter Elst at the La Silla Observatory in Chile on July 11, 1994.[1]

See also

- List of minor planets: 152001–153000

References

[1] JPL Small-Body Database Browser (http://ssd.jpl.nasa.gov/sbdb.cgi?sstr=152576)

Minor planet

A **minor planet** is an astronomical object in direct orbit around the Sun that is neither a dominant planet nor originally classified as a comet. Minor planets can be dwarf planets, asteroids, trojans, centaurs, Kuiper belt objects, and other trans-Neptunian objects.[1] The first minor planet discovered was Ceres in 1801 (although from the time of its discovery until 1851 it was considered to be a planet). The orbits of more than 570,000 objects have been archived at the Minor Planet Center.[2]

The term "minor planet" has been used since the 19th century to describe these objects.[3] The term **planetoid** has also been used, especially for larger (planetary) objects.[4] Historically, the terms *asteroid*, *minor planet*, and *planetoid* have been more or less synonymous,[4] [5] but the issue has been complicated by the discovery of numerous minor planets beyond the orbit of Jupiter and especially Neptune that are not universally considered asteroids.[5] Minor planets seen outgassing may receive a dual classification as a comet.

Before 2006 the International Astronomical Union had officially used the term *minor planet*. During its 2006 meeting, the Union reclassified minor planets and comets into dwarf planets and small Solar System bodies.[6] Objects are called dwarf planets if their self-gravity is sufficient to achieve hydrostatic equilibrium, that is, an ellipsoidal shape, with all other minor planets and comets called "small Solar System bodies".[6] The IAU states: "the term 'minor planet' may still be used, but generally the term 'small solar system body' will be preferred."[7] However, for purposes of numbering and naming, the traditional distinction between minor planet and comet is still followed.

Populations

Hundreds of thousands of minor planets have been discovered within the Solar System, with the 2009 rate of discovery at over 3,000 per month. Of the more than 535,000 registered minor planets, 251,651 have orbits known well enough to be assigned permanent official numbers.[8] [2] Of these, 16,154 have official names[2] . As of September 2010, the lowest-numbered unnamed minor planet is (3708) 1974 FV_1;[9] but there are also some named minor planets above number 240,000.[10]

There are various broad minor-planet populations:

- Asteroids; traditionally, most have been bodies in the inner Solar System.[5]
  - Main-belt asteroids, those following roughly circular orbits between Mars and Jupiter. These are the original and best-known group of asteroids or minor planets.
  - Near-Earth asteroids, those whose orbits take them inside the orbit of Mars. Further subclassification of these, based on orbital distance, is used:[11]

        - Aten asteroids, those that have semi-major axes of less than one Earth orbit. Those Aten asteroids that have their aphelion within Earth's orbit are known as Apohele asteroids;
        - Amor asteroids are those near-Earth asteroids that approach the orbit of the Earth from beyond, but do not cross it. Amor asteroids are further subdivided into four subgroups, depending on where their semimajor axis falls between Earth's orbit and the asteroid belt;
        - Apollo asteroids are those asteroids with a semimajor axis greater than Earth's, while having a perihelion distance of 1.017 AU or less. Like Aten asteroids, Apollo asteroids are Earth-crossers.
    - Earth trojans, asteroids sharing Earth's orbit and gravitationally locked to it. As of 2011, the only one known is 2010 TK_7.[12]
    - Mars trojans, asteroids sharing Mars's orbit and gravitationally locked to it. As of 2007, eight such asteroids are known.[13]
    - Jupiter trojans, asteroids sharing Jupiter's orbit and gravitationally locked to it. Numerically they are estimated to equal the main-belt asteroids.
- Distant minor planets; an umbrella term for minor planets in the outer Solar System.
    - Centaurs, bodies in the outer Solar System between Jupiter and Neptune. They have unstable orbits due to the gravitational influence of the giant planets, and therefore must have come from elsewhere, probably outside Neptune.[14]
    - Neptune trojans, bodies sharing Neptune's orbit and gravitationally locked to it. Although only a handful are known, there is evidence that Neptune trojans are more numerous than either the asteroids in the asteroid belt or the Jupiter trojans.[15]
    - Trans-Neptunian objects, bodies at or beyond the orbit of Neptune, the outermost planet.
        - The Kuiper belt, objects inside an apparent population drop-off approximately 55 AU from the Sun.
            - Classical Kuiper belt objects, also known as cubewanos, are in primordial, relatively circular orbits that are not in resonance with Neptune.
            - Plutinos, bodies like Pluto that are in a 2:3 resonance with Neptune, and other resonant trans-Neptunian objects. This is a rather large group.
        - Scattered disc, objects with aphelia outside the Kuiper belt. These are thought to have been scattered by Neptune.
        - Detached objects such as Sedna, with both aphelia and perihelia outside the Kuiper belt.
        - The Oort Cloud, a hypothetical population thought to be the source of long-period comets that may extend out to 50,000 AU from the Sun.

Naming

A newly discovered minor planet is given a provisional designation (such as 2002 AT_4) consisting of the year of discovery and an alphanumeric code indicating the half-month of discovery and the sequence within that half-month. Once an asteroid's orbit has been confirmed, it is given a number, and later may also be given a name (e.g. 433 Eros). The formal naming convention uses parentheses around the number, but dropping the parentheses is quite common. Informally, it is common to drop the number altogether, or to drop it after the first mention when a name is repeated in running text.

Minor planets that have been given a number but not a name keep their provisional designation, e.g. (29075) 1950 DA. As modern discovery techniques are finding vast numbers of new asteroids, they are increasingly being left unnamed. The earliest discovered to be left unnamed was for a long time (3360) 1981 VA, now 3360 Syrinx; as of September 2008, this distinction is held by (3708) 1974 FV_1. On rare occasions, a small object's provisional designation may become used as a name in itself: the still unnamed (15760) 1992 QB_1 gave its "name" to a group of Kuiper belt objects which became known as cubewanos.

A few objects are cross-listed as both comets and asteroids, such as 4015 Wilson–Harrington, which is also listed as 107P/Wilson–Harrington.

Numbering

Minor planets are awarded an official number once their orbits are confirmed. With the increasing rapidity of discovery, these are now six-figure numbers. The switch from five figures to six figures arrived with the publication of the Minor Planet Circular (MPC) of October 19, 2005, which saw the highest numbered minor planet jump from 99947 to 118161.

Sources for names

The first few asteroids were named after figures from Greek and Roman mythology, but as such names started to dwindle the names of famous people, literary characters, discoverer's wives, children, and even television characters were used.

The first asteroid to be given a non-mythological name was 20 Massalia, named after the Greek name for the city of Marseilles. The first to be given an entirely non-Classical name was 45 Eugenia, named after Empress Eugénie de Montijo, the wife of Napoleon III. For some time only female (or feminized) names were used; Alexander von Humboldt was the first man to have an asteroid named after him, but his name was feminized to 54 Alexandra. This unspoken tradition lasted until 334 Chicago was named; even then, oddly feminised names show up in the list for years after.

As the number of asteroids began to run into the hundreds, and eventually the thousands, discoverers began to give them increasingly frivolous names. The first hints of this were 482 Petrina and 483 Seppina, named after the discoverer's pet dogs. However, there was little controversy about this until 1971, upon the naming of 2309 Mr. Spock (the name of the discoverer's cat). Although the IAU subsequently banned pet names as sources, eccentric asteroid names are still being proposed and accepted, such as 4321 Zero, 6042 Cheshirecat, 9007 James Bond, 13579 Allodd and 24680 Alleven, and 26858 Misterrogers.

A well-established rule is that, unlike comets, minor planets may not be named after their discoverer(s). One way to circumvent this rule has been for astronomers to exchange the courtesy of naming their discoveries after each other. An exception to this rule is 96747 Crespodasilva, which was named after its discoverer, Lucy d'Escoffier Crespo da Silva, because she died shortly after the discovery, at age 22.[16] [17]

Names were adapted to various languages from the beginning. 1 Ceres, *Ceres* being its Anglo-Latin name, was actually named *Cerere*, the Italian form of the name. German, French, Arabic and Hindi use forms similar to the English, whereas Russian uses a form, *Tserera*, similar to the Italian. In Greek the name was translated to Demeter, the Greek equivalent of the Roman goddess Ceres. In the early years, before it started causing conflicts, asteroids named after Roman figures were generally translated in Greek; other examples are Hera for 3 Juno, Hestia for 4 Vesta, Chloris for 8 Flora, and Pistê for 37 Fides. In Chinese, the names are not given the Chinese forms of the deities they are named after, but rather typically have a syllable or two for the character of the deity or person, followed by 'god(dess)' or 'woman' if just one syllable, plus 'star/planet', so that most asteroid names are written with three Chinese characters. Thus Ceres is 'grain goddess planet',[18] Pallas is 'wisdom goddess planet', etc.

Special naming rules

Minor-planet naming is not always a free-for-all: there are some populations for which rules have developed about the sources of names. For instance, centaurs (orbiting between Saturn and Neptune) are all named after mythological centaurs; Jupiter trojans after heroes from the Trojan War; resonant trans-Neptunian objects after underworld spirits; and non-resonant TNOs after creation deities.

Physical properties of comets and minor planets

Commission 15[19] of the International Astronomical Union is dedicated to the Physical Study of Comets & Minor Planets.

Archival data on the physical properties of comets and minor planets are found in the PDS Asteroid/Dust Archive.[20] This includes standard asteroid physical characteristics such as the properties of binary systems, occultation timings and diameters, masses, densities, rotation periods, surface temperatures, albedoes, spin vectors, taxonomy, and absolute magnitudes and slopes. In addition, European Asteroid Research Node (E.A.R.N.), an association of asteroid research groups, maintains a Data Base of Physical and Dynamical Properties of Near Earth Asteroids.[21]

Most detailed information is available from Category: Asteroids visited by spacecraft and Category: Comets visited by spacecraft.

See also

- List of minor planets
- Solar System
- Groups of minor planets
- Small Solar System body
- Plutoid
- Quasi-satellite

External links

- Minor Planet Center [22]

References

[1] "Unusual Minor Planets" (http://www.minorplanetcenter.net/iau/lists/Unusual.html). Minor Planet Center. . Retrieved 23 December 2011.

[2] "Minor Planet Statistics" (http://www.minorplanetcenter.net/iau/lists/ArchiveStatistics.html). Minor Planet Center. . Retrieved 2011-02-26.

[3] When did the asteroids become minor planets? (http://aa.usno.navy.mil/faq/docs/minorplanets.php), James L. Hilton, Astronomical Information Center, United States Naval Observatory. Accessed May 5, 2008.

[4] Planet, asteroid, minor planet: A case study in astronomical nomenclature, David W. Hughes, Brian G. Marsden, *Journal of Astronomical History and Heritage* **10**, #1 (2007), pp. 21–30. Bibcode: 2007JAHH...10...21H

[5] " Asteroid (http://encarta.msn.com/encyclopedia_761551567/asteroid.html)", *MSN Encarta*, Microsoft. Accessed May 5, 2008. Archived (http://www.webcitation.org/5kx3nDxZi) 2009-11-01.

[6] Press release, IAU 2006 General Assembly: Result of the IAU Resolution votes (http://www.iau.org/public_press/news/detail/iau0603/), International Astronomical Union, August 24, 2006. Accessed May 5, 2008.

[7] Questions and Answers on Planets (http://www.iau.org/public_press/news/release/iau0603/questions_answers/), additional information, news release IAU0603, IAU 2006 General Assembly: Result of the IAU Resolution votes, International Astronomical Union, August 24, 2006. Accessed May 8, 2008.

[8] JPL. "How Many Solar System Bodies" (http://ssd.jpl.nasa.gov/?body_count). *JPL Solar System Dynamics*. NASA. . Retrieved 2010-09-27.

[9] "Discovery Circumstances: Numbered Minor Planets (1)-(5000)" (http://www.minorplanetcenter.net/iau/lists/NumberedMPs000001.html). Minor Planet Center. . Retrieved 2011-02-26.
[10] "Discovery Circumstances: Numbered Minor Planets (240001)-(245000)" (http://www.minorplanetcenter.net/iau/lists/NumberedMPs240001.html). Minor Planet Center. . Retrieved 2011-02-26.
[11] Yeomans, Don, "Near-Earth Object groups" (http://neo.jpl.nasa.gov/neo/groups.html), *Near Earth Object Project* (NASA), , retrieved 2011-12-24
[12] Connors, Martin; Wiegert, Paul; Veillet, Christian (July 2011), "Earth's Trojan asteroid", *Nature* **475** (7357): 481–483, Bibcode 2011Natur.475..481C, doi:10.1038/nature10233, PMID 21796207
[13] Trilling, David et al. (October 2007), "DDT observations of five Mars Trojan asteroids", *Spitzer Proposal ID #465*, Bibcode 2007sptz.prop..465T
[14] Horner, J.; Evans, N.W.; Bailey, M. E. (2004). "Simulations of the Population of Centaurs I: The Bulk Statistics". *Monthly Notices of the Royal Astronomical Society* **354** (3): 798–810. arXiv:astro-ph/0407400. Bibcode 2004MNRAS.354..798H. doi:10.1111/j.1365-2966.2004.08240.x.
[15] Neptune trojans, Jupiter trojans (http://www.dtm.ciw.edu/users/sheppard/trojans/)
[16] NASA JPL Small-Body Database Browser on 96747 Crespodasilva (http://ssd.jpl.nasa.gov/sbdb.cgi?sstr=96747+Crespodasilva)
[17] Staff (November 28, 2000). "Lucy Crespo da Silva, 22, a senior, dies in fall" (http://hubblesite.org/newscenter/newsdesk/archive/releases/1991/12/text/). Hubble News Desk. . Retrieved 2008-04-15.
[18] 'valley' being a common abbreviation of 'grain' that would be formally adopted with simplified Chinese characters.
[19] "Division III Commission 15 Physical Study of Comets & Minor Planets" (http://www.iau.org/science/scientific_bodies/commissions/15). International Astronomical Union (IAU). September 29, 2005. . Retrieved 2010-03-22.
[20] "Physical Properties of Asteroids" (http://www.psi.edu/pds/archive/physical.html). .
[21] "The Near-Earth Asteroids Data Base" (http://earn.dlr.de/nea). .
[22] http://www.minorplanetcenter.net/iau/mpc.html

Asteroid belt

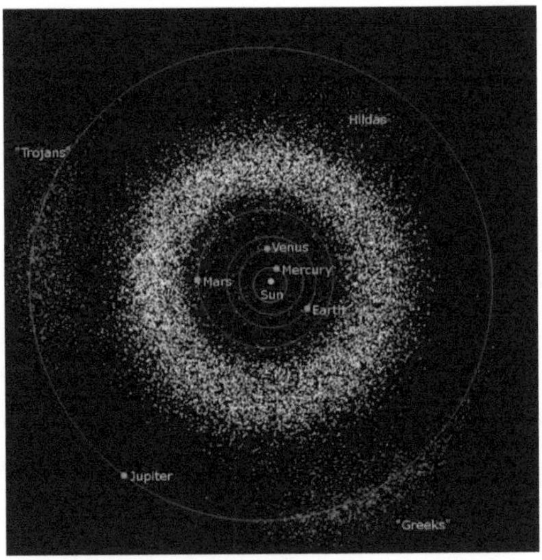

The asteroid belt (shown in white) is located between the orbits of Mars and Jupiter.

The **asteroid belt** is the region of the Solar System located roughly between the orbits of the planets Mars and Jupiter. It is occupied by numerous irregularly shaped bodies called asteroids or minor planets. The asteroid belt is also termed the **main asteroid belt** or **main belt** to distinguish its members from other asteroids in the Solar System such as near-Earth asteroids and trojan asteroids. About half the mass of the belt is contained in the four largest asteroids, Ceres, Vesta, Pallas, and Hygiea. These have mean diameters of more than 400 km, while Ceres, the asteroid belt's only dwarf planet, is about 950 km in diameter.[1] [2] [3] [4] The remaining bodies range down to the size of a dust particle. The asteroid material is so thinly distributed that numerous unmanned spacecraft have traversed it without incident. Nonetheless, collisions between large asteroids do occur, and these can form an asteroid family whose members have similar orbital characteristics and compositions. Nesvorny and Jenniskens (2010 Astrophysical Journal) attributed 85 percent of the Zodiacal Light dust to fragmentations of

Jupiter-family comets, rather than from comets and collisions between asteroids in the asteroid belt. It was once thought that collisions of asteroids produce a fine dust that forms a major component of the zodiacal light. Individual asteroids within the asteroid belt are categorized by their spectra, with most falling into three basic groups: carbonaceous (C-type), silicate (S-type), and metal-rich (M-type).

The asteroid belt formed from the primordial solar nebula as a group of planetesimals, the smaller precursors of the planets, which in turn formed protoplanets. Between Mars and Jupiter, however, gravitational perturbations from the giant planet imbued the protoplanets with too much orbital energy for them to accrete into a planet. Collisions became too violent, and instead of fusing together, the planetesimals and most of the protoplanets shattered. As a result, most of the asteroid belt's mass has been lost since the formation of the Solar System. Some fragments can eventually find their way into the inner Solar System, leading to meteorite impacts with the inner planets. Asteroid orbits continue to be appreciably perturbed whenever their period of revolution about the Sun forms an orbital resonance with Jupiter. At these orbital distances, a Kirkwood gap occurs as they are swept into other orbits.

Classes of small Solar System bodies in other regions include the centaurs, Kuiper belt and scattered disk objects, and Oort cloud comets.

History of observation

Giuseppe Piazzi, discoverer of Ceres, known as a planet for many years, then as asteroid number 1, and eventually, a dwarf planet

In an anonymous footnote to his 1766 translation of Charles Bonnet's *Contemplation de la Nature*,[5] the astronomer Johann Daniel Titius of Wittenberg[6] [7] noted an apparent pattern in the layout of the planets. If one began a numerical sequence at 0, then included 3, 6, 12, 24, 48, etc., doubling each time, and added four to each number and divided by 10, this produced a remarkably close approximation to the radii of the orbits of the known planets as measured in astronomical units. This pattern, now known as the Titius–Bode law, predicted the semi-major axes of the six planets of the time (Mercury, Venus, Earth, Mars, Jupiter and Saturn) provided one allowed for a "gap" between the orbits of Mars and Jupiter. In his footnote Titius declared, "But should the Lord Architect have left that space empty? Not at all."[6] In 1768, the astronomer Johann Elert Bode made note of Titius's relationship in his *Anleitung zur Kenntniss des gestirnten Himmels* (English: *Instruction for the Knowledge of the Starry Heavens*) but did not credit Titius until later editions. It became known as "Bode's law".[7] When William Herschel discovered Uranus in 1781, the planet's orbit matched the law almost perfectly, leading astronomers to conclude that there had to be a planet between the orbits of Mars and Jupiter.

In 1800 the astronomer Baron Franz Xaver von Zach recruited 24 of his fellows into a club, the Vereinigte Astronomische Gesellschaft ("United Astronomical Society") which he informally dubbed the "Lilienthal Society"[8] for its meetings in Lilienthal, a small city near Bremen. Determined to bring the Solar System to order, the group became known as the "Himmelspolizei", or Celestial Police. Notable members included Herschel, the British Astronomer Royal Nevil Maskelyne, Charles Messier, and Heinrich Olbers.[9] The Society assigned to each astronomer a 15° region of the zodiac to search for the missing planet.[10]

Only a few months later, a non-member of the Celestial Police confirmed their expectations. On January 1, 1801, Giuseppe Piazzi, Chair of Astronomy at the University of Palermo, Sicily, found a tiny moving object in an orbit with exactly the radius predicted by the Titius–Bode law. He dubbed it Ceres, after the Roman goddess of the harvest and patron of Sicily. Piazzi initially believed it a comet, but its lack of a coma suggested it was a planet.[9] Fifteen months later, Heinrich Wilhelm Olbers discovered a second object in the same region, Pallas. Unlike the other known planets, the objects remained points of light even under the highest telescope magnifications, rather than resolving into discs. Apart from their rapid movement, they appeared indistinguishable from stars. Accordingly, in 1802 William Herschel suggested they be placed into a separate category, named *asteroids*, after the Greek *asteroeides*, meaning "star-like".[11] [12] Upon completing a series of observations of Ceres and Pallas, he concluded,[13]

> Neither the appellation of planets, nor that of comets, can with any propriety of language be given to these two stars ... They resemble small stars so much as hardly to be distinguished from them. From this, their asteroidal appearance, if I take my name, and call them Asteroids; reserving for myself however the liberty of changing that name, if another, more expressive of their nature, should occur.

Despite Herschel's coinage, for several decades it remained common practice to refer to these objects as planets.[5] By 1807, further investigation revealed two new objects in the region: 3 Juno and 4 Vesta.[14] The burning of Lilienthal in the Napoleonic wars brought this first period of discovery to a close,[14] and only in 1845 did astronomers detect another object (5 Astraea). Shortly thereafter new objects were found at an accelerating rate, and counting them among the planets became increasingly cumbersome. Eventually, they were dropped from the planet list as first suggested by Alexander von Humboldt in the early 1850s, and William Herschel's choice of nomenclature, "asteroids", gradually came into common use.[5]

The discovery of Neptune in 1846 led to the discrediting of the Titius–Bode law in the eyes of scientists, as its orbit was nowhere near the predicted position. To date, there is no scientific explanation for the law, and astronomers' consensus regards it as a coincidence.[15]

The expression "asteroid belt" came into use in the very early 1850s, although it is hard to pinpoint who coined the term. The first English use seems to be in the 1850 translation (by E. C. Otté) of Alexander von Humboldt's *Cosmos:*[16] "[...] and the regular appearance, about the 13th of November and the 11th of August, of shooting stars, which probably form part of a belt of asteroids intersecting the Earth's orbit and moving with planetary velocity". Other early appearances occur in Robert James Mann's *A Guide to the Knowledge of the Heavens,*[17] "The orbits of the asteroids are placed in a wide belt of space, extending between the extremes of [...]". The American astronomer Benjamin Peirce seems to have adopted that terminology and to have been one of its promoters.[18] One hundred asteroids had been located by mid-1868, and in 1891 the introduction of astrophotography by Max Wolf accelerated the rate of discovery still further.[19] A total of 1,000 asteroids had been found by 1921,[20] 10,000 by 1981,[21] and 100,000 by 2000.[22] Modern asteroid survey systems now use automated means to locate new minor planets in ever-increasing quantities.

Origin

Formation

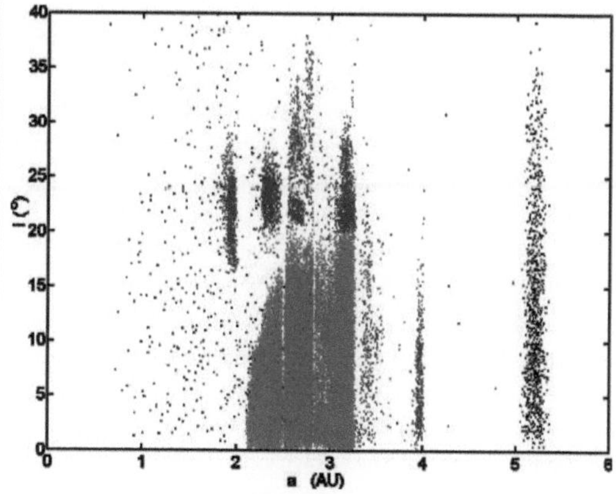

The asteroid belt showing the orbital inclinations versus distances from the Sun, with asteroids in the core region of the asteroid belt in red and other asteroids in blue

In 1802, shortly after discovering Pallas, Heinrich Olbers suggested to William Herschel that Ceres and Pallas were fragments of a much larger planet that once occupied the Mars–Jupiter region, this planet having suffered an internal explosion or a cometary impact many million years before.[23] Over time, however, this hypothesis has fallen from favor. The large amount of energy that would have been required to destroy a planet, combined with the belt's low combined mass, which is only about 4% of the mass of the Earth's Moon,[1] do not support the hypothesis. Further, the significant chemical differences between the asteroids are difficult to explain if they come from the same planet.[24] Today, most scientists accept that, rather than fragmenting from a progenitor planet, the asteroids never formed a planet at all.

In general in the Solar System, planetary formation is thought to have occurred via a process comparable to the long-standing nebular hypothesis: a cloud of interstellar dust and gas collapsed under the influence of gravity to form a rotating disk of material that then further condensed to form the Sun and planets.[25] During the first few million years of the Solar System's history, an accretion process of sticky collisions caused the clumping of small particles, which gradually increased in size. Once the clumps reached sufficient mass, they could draw in other bodies through gravitational attraction and become planetesimals. This gravitational accretion led to the formation of the rocky planets and the gas giants.

Planetesimals within the region which would become the asteroid belt were too strongly perturbed by Jupiter's gravity to form a planet. Instead they continued to orbit the Sun as before, while occasionally colliding.[26] In regions where the average velocity of the collisions was too high, the shattering of planetesimals tended to dominate over accretion,[27] preventing the formation of planet-sized bodies. Orbital resonances occurred where the orbital period of an object in the belt formed an integer fraction of the orbital period of Jupiter, perturbing the object into a different orbit; the region lying between the orbits of Mars and Jupiter contains many such orbital resonances. As Jupiter migrated inward following its formation, these resonances would have swept across the asteroid belt, dynamically exciting the region's population and increasing their velocities relative to each other.[28]

During the early history of the Solar System, the asteroids melted to some degree, allowing elements within them to be partially or completely differentiated by mass. Some of the progenitor bodies may even have undergone periods of explosive volcanism and formed magma oceans. However, because of the relatively small size of the bodies, the period of melting was necessarily brief (compared to the much larger planets), and had generally ended about 4.5 billion years ago, in the first tens of millions of years of formation.[29] In August 2007, a study of zircon crystals in an Antarctic meteorite believed to have originated from 4 Vesta suggested that it, and by extension the rest of the asteroid belt, had formed rather quickly, within ten million years of the Solar System's origin.[30]

Evolution

The asteroids are not samples of the primordial Solar System. They have undergone considerable evolution since their formation, including internal heating (in the first few tens of millions of years), surface melting from impacts, space weathering from radiation, and bombardment by micrometeorites.[31] While some scientists refer to the asteroids as residual planetesimals,[32] other scientists consider them distinct.[33]

The current asteroid belt is believed to contain only a small fraction of the mass of the primordial belt. Computer simulations suggest that the original asteroid belt may have contained mass equivalent to the Earth. Primarily because of gravitational perturbations, most of the material was ejected from the belt within about a million years of formation, leaving behind less than 0.1% of the original mass.[26] Since their formation, the size distribution of the asteroid belt has remained relatively stable: there has been no significant increase or decrease in the typical dimensions of the main-belt asteroids.[34]

The 4:1 orbital resonance with Jupiter, at a radius 2.06 AU, can be considered the inner boundary of the asteroid belt. Perturbations by Jupiter send bodies straying there into unstable orbits. Most bodies formed inside the radius of this gap were swept up by Mars (which has an aphelion at 1.67 AU) or ejected by its gravitational perturbations in the early history of the Solar System.[35] The Hungaria asteroids lie closer to the Sun than the 4:1 resonance, but are protected from disruption by their high inclination.[36]

When the asteroid belt was first formed, the temperatures at a distance of 2.7 AU from the Sun formed a "snow line" below the freezing point of water. Planetesimals formed beyond this radius were able to accumulate ice.[37] [38] In 2006 it was announced that a population of comets had been discovered within the asteroid belt beyond the snow line, which may have provided a source of water for Earth's oceans. According to some models, there was insufficient outgassing of water during the Earth's formative period to form the oceans, requiring an external source such as a cometary bombardment.[39]

Characteristics

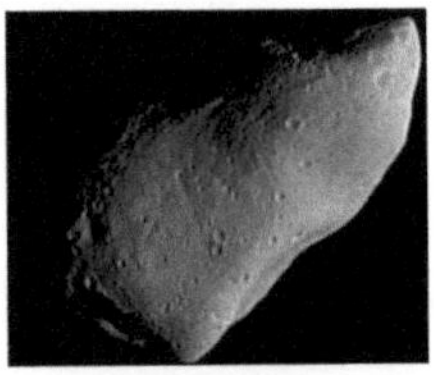

951 Gaspra, the first asteroid imaged by a spacecraft, as viewed during *Galileo*'s 1991 flyby; colors are exaggerated

Fragment of the Allende meteorite, a carbonaceous chondrite that fell to Earth in Mexico in 1969

Contrary to popular imagery, the asteroid belt is mostly empty. The asteroids are spread over such a large volume that it would be improbable to reach an asteroid without aiming carefully. Nonetheless, hundreds of thousands of asteroids are currently known, and the total number ranges in the millions or more, depending on the lower size cutoff. Over 200 asteroids are known to be larger than 100 km,[40] while a survey in the infrared wavelengths shows that the asteroid belt has 700,000 to 1.7 million asteroids with a diameter of 1 km or more.[41] The apparent magnitudes of most of the known asteroids are 11–19, with the median at about 16.[42]

The total mass of the asteroid belt is estimated to be 2.8×10^{21} to 3.2×10^{21} kilograms, which is just 4% of the mass of the Moon.[2] The four largest objects, Ceres, 4 Vesta, 2 Pallas, and 10 Hygiea, account for half of the belt's total mass, with almost one-third accounted for by Ceres alone.[3] [4]

Composition

The current belt consists primarily of three categories of asteroids: C-type or carbonaceous asteroids, S-type or silicate asteroids, and M-type or metallic asteroids.

Carbonaceous asteroids, as their name suggests, are carbon-rich and dominate the belt's outer regions.[43] Together they comprise over 75% of the visible asteroids. They are more red in hue than the other asteroids and have a very low albedo. Their surface composition is similar to carbonaceous chondrite meteorites. Chemically, their spectra match the primordial composition of the early Solar System, with only the lighter elements and volatiles removed.

S-type (silicate-rich) asteroids are more common toward the inner region of the belt, within 2.5 AU of the Sun.[43] [44] The spectra of their surfaces reveal the presence of silicates and some metal, but no significant carbonaceous compounds. This indicates that their materials have been significantly modified from their primordial composition, probably through melting and reformation. They have a relatively high albedo, and form about 17% of the total asteroid population.

M-type (metal-rich) asteroids form about 10% of the total population; their spectra resemble that of iron-nickel. Some are believed to have formed from the metallic cores of differentiated progenitor bodies that were disrupted through collision. However, there are also some silicate compounds that can produce a similar appearance. For example, the large M-type asteroid 22 Kalliope does not appear to be primarily composed of metal.[45] Within the asteroid belt, the number distribution of M-type asteroids peaks at a semi-major axis of about 2.7 AU.[46] It is not yet clear whether all M-types are compositionally similar, or whether it is a label for several varieties which do not fit neatly into the main C and S classes.[47]

One mystery of the asteroid belt is the relative rarity of V-type, or basaltic asteroids.[48] Theories of asteroid formation predict that objects the size of Vesta or larger should form crusts and mantles, which would be composed mainly of basaltic rock, resulting in more than half of all asteroids being composed either of basalt or olivine. Observations, however, suggest that 99 percent of the predicted basaltic material is missing.[49] Until 2001, most basaltic bodies discovered in the asteroid belt were believed to originate from the asteroid Vesta (hence their name V-type). However, the discovery of the asteroid 1459 Magnya revealed a slightly different chemical composition from the other basaltic asteroids discovered until then, suggesting a different origin.[49] This hypothesis was reinforced by the further discovery in 2007 of two asteroids in the outer belt, 7472 Kumakiri and (10537) 1991 RY_{16}, with differing basaltic composition that could not have originated from Vesta. These latter two are the only V-type asteroids discovered in the outer belt to date.[48]

The temperature of the asteroid belt varies with the distance from the Sun. For dust particles within the belt, typical temperatures range from 200 K (−73 °C) at 2.2 AU down to 165 K (−108 °C) at 3.2 AU[50] However, due to rotation, the surface temperature of an asteroid can vary considerably as the sides are alternately exposed to solar radiation and then to the stellar background.

Main-belt comets

Several otherwise unremarkable bodies in the outer belt show cometary activity. Since their orbits cannot be explained through capture of classical comets, it is thought that many of the outer asteroids may be icy, with the ice occasionally exposed to sublimation through small impacts. Main-belt comets may have been a major source of the Earth's oceans, since the deuterium-hydrogen ratio is too low for classical comets to have been the principal source.[51]

Orbits

Most asteroids within the asteroid belt have orbital eccentricities of less than 0.4, and an inclination of less than 30°. The orbital distribution of the asteroids reaches a maximum at an eccentricity of around 0.07 and an inclination below 4°.[42] Thus while a typical asteroid has a relatively circular orbit and lies near the plane of the ecliptic, some asteroid orbits can be highly eccentric or travel well outside the ecliptic plane.

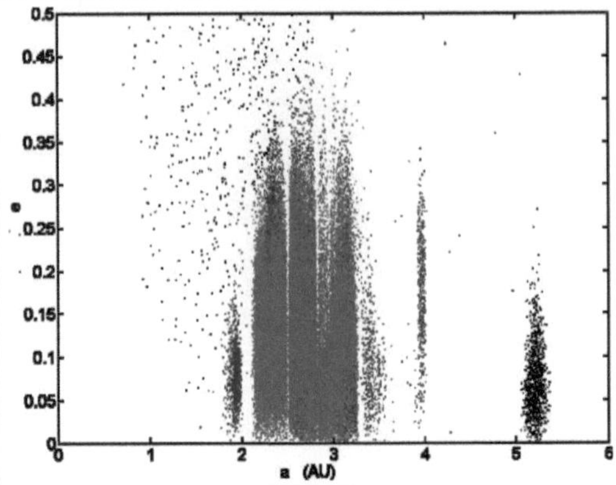

The asteroid belt (showing eccentricities), with the asteroid belt in red and blue ("core" region in red)

Sometimes, the term *main belt* is used to refer only to the more compact "core" region where the greatest concentration of bodies is found. This lies between the strong 4:1 and 2:1 Kirkwood gaps at 2.06 and 3.27 AU, and at orbital eccentricities less than roughly 0.33, along with orbital inclinations below about 20°. This "core" region contains approximately 93.4% of all numbered minor planets within the Solar System.[52]

Kirkwood gaps

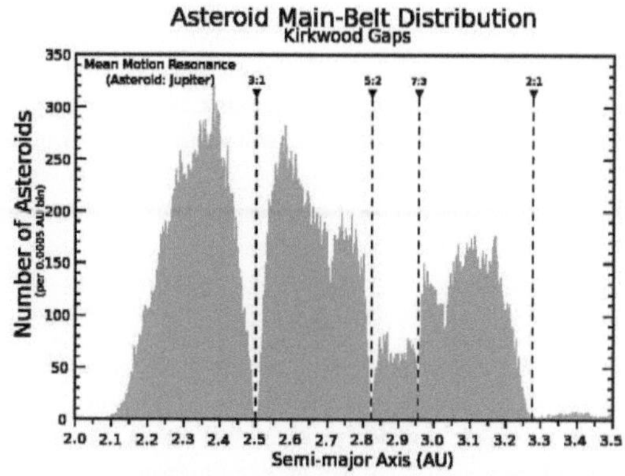

This chart shows the distribution of asteroid semi-major axes in the "core" of the asteroid belt. Black arrows point to the Kirkwood gaps, where orbital resonances with Jupiter destabilize orbits.

The semi-major axis of an asteroid is used to describe the dimensions of its orbit around the Sun, and its value determines the minor planet's orbital period. In 1866, Daniel Kirkwood announced the discovery of gaps in the distances of these bodies' orbits from the Sun. They were located at positions where their period of revolution about the Sun was an integer fraction of Jupiter's orbital period. Kirkwood proposed that the gravitational perturbations of the planet led to the removal of asteroids from these orbits.[53]

When the mean orbital period of an asteroid is an integer fraction of the orbital period of Jupiter, a mean-motion resonance with the gas giant is created that is sufficient to perturb an asteroid to new orbital elements. Asteroids that become located in the gap orbits (either primordially because of the migration of Jupiter's orbit,[54] or due to prior perturbations or collisions) are gradually nudged into different, random orbits with a larger or smaller semi-major axis.

The gaps are not seen in a simple snapshot of the locations of the asteroids at any one time because asteroid orbits are elliptical, and many asteroids still cross through the radii corresponding to the gaps. The actual spatial density of asteroids in these gaps does not differ significantly from the neighboring regions.[55]

The main gaps occur at the 3:1, 5:2, 7:3, and 2:1 mean-motion resonances with Jupiter. An asteroid in the 3:1 Kirkwood gap would orbit the Sun three times for each Jovian orbit, for instance. Weaker resonances occur at other semi-major axis values, with fewer asteroids found than nearby. (For example, an 8:3 resonance for asteroids with a semi-major axis of 2.71 AU.)[56]

The main or core population of the asteroid belt is sometimes divided into three zones, based on the most prominent Kirkwood gaps. Zone I lies between the 4:1 resonance (2.06 AU) and 3:1 resonance (2.5 AU) Kirkwood gaps. Zone II continues from the end of Zone I out to the 5:2 resonance gap (2.82 AU). Zone III extends from the outer edge of Zone II to the 2:1 resonance gap (3.28 AU).[57]

The asteroid belt may also be divided into the inner and outer belts, with the inner belt formed by asteroids orbiting nearer to Mars than the 3:1 Kirkwood gap (2.5 AU), and the outer belt formed by those asteroids closer to Jupiter's orbit. (Some authors subdivide the inner and outer belts at the 2:1 resonance gap (3.3 AU), while others suggest inner, middle, and outer belts.)

Collisions

The zodiacal light, created in part by dust from collisions in the asteroid belt

The high population of the asteroid belt makes for a very active environment, where collisions between asteroids occur frequently (on astronomical time scales). Collisions between main-belt bodies with a mean radius of 10 km are expected to occur about once every 10 million years.[58] A collision may fragment an asteroid into numerous smaller pieces (leading to the formation of a new asteroid family). Conversely, collisions that occur at low relative speeds may also join two asteroids. After more than 4 billion years of such processes, the members of the asteroid belt now bear little resemblance to the original population.

Along with the asteroid bodies, the asteroid belt also contains bands of dust with particle radii of up to a few hundred micrometres. This fine material is produced, at least in part, from collisions between asteroids, and by the impact of micrometeorites upon the asteroids. Due to the Poynting-Robertson effect, the pressure of solar radiation causes this dust to slowly spiral inward toward the Sun.[59]

The combination of this fine asteroid dust, as well as ejected cometary material, produces the zodiacal light. This faint auroral glow can be viewed at night extending from the direction of the Sun along the plane of the ecliptic. Particles that produce the visible zodiacal light average about 40 μm in radius. The typical lifetimes of such particles are about 700,000 years. Thus, to maintain the bands of dust, new particles must be steadily produced within the asteroid belt.[59]

Meteorites

Some of the debris from collisions can form meteoroids that enter the Earth's atmosphere.[60] Of the 50,000 meteorites found on Earth to date, 99.8 percent are believed to have originated in the asteroid belt.[61] A September 2007 study has suggested that a large-body collision undergone by the asteroid 298 Baptistina sent a number of fragments into the inner Solar System. The impacts of these fragments are believed to have created both Tycho crater on the Moon and Chicxulub crater in Mexico, the relict of the massive impact which is believed to have triggered the extinction of the dinosaurs 65 million years ago.[62]

Families and groups

In 1918, the Japanese astronomer Kiyotsugu Hirayama noticed that the orbits of some of the asteroids had similar parameters, forming families or groups.[63]

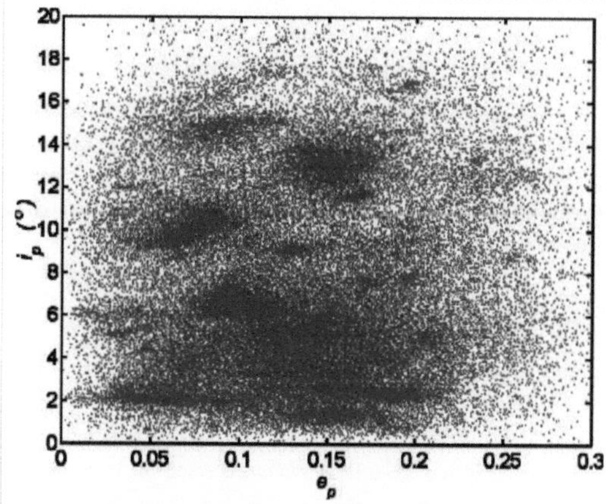

This plot of orbital inclination (i_p) versus eccentricity (e_p) for the numbered main-belt asteroids clearly shows clumpings representing asteroid families.

Approximately one-third of the asteroids in the asteroid belt are members of an asteroid family. These share similar orbital elements, such as semi-major axis, eccentricity, and orbital inclination as well as similar spectral features, all of which indicate a common origin in the breakup of a larger body. Graphical displays of these elements, for members of the asteroid belt, show concentrations indicating the presence of an asteroid family. There are about 20–30 associations that are almost certainly asteroid families. Additional groupings have been found that are less certain. Asteroid families can be confirmed when the members display common spectral features.[64] Smaller associations of asteroids are called groups or clusters.

Some of the most prominent families in the asteroid belt (in order of increasing semi-major axes) are the Flora, Eunoma, Koronis, Eos, and Themis families.[46] The Flora family, one of the largest with more than 800 known members, may have formed from a collision less than a billion years ago.[65] The largest asteroid to be a true member of a family (as opposed to an interloper in the case of Ceres with the Gefion family) is 4 Vesta. The Vesta family is believed to have formed as the result of a crater-forming impact on Vesta. Likewise, the HED meteorites may also have originated from Vesta as a result of this collision.[66]

Three prominent bands of dust have been found within the asteroid belt. These have similar orbital inclinations as the Eos, Koronis, and Themis asteroid families, and so are possibly associated with those groupings.[67]

Periphery

Skirting the inner edge of the belt (ranging between 1.78 and 2.0 AU, with a mean semi-major axis of 1.9 AU) is the Hungaria family of minor planets. They are named after the main member, 434 Hungaria; the group contains at least 52 named asteroids. The Hungaria group is separated from the main body by the 4:1 Kirkwood gap and their orbits have a high inclination. Some members belong to the Mars-crossing category of asteroids, and gravitational perturbations by Mars are likely a factor in reducing the total population of this group.[68]

Another high-inclination group in the inner part of the asteroid belt is the Phocaea family. These are composed primarily of S-type asteroids, whereas the neighboring Hungaria family includes some E-types.[69] The Phocaea family orbit between 2.25 and 2.5 AU from the Sun.

Skirting the outer edge of the asteroid belt is the Cybele group, orbiting between 3.3 and 3.5 AU. These have a 7:4 orbital resonance with Jupiter. The Hilda family orbit between 3.5 and 4.2 AU, and have relatively circular orbits and a stable 3:2 orbital resonance with Jupiter. There are few asteroids beyond 4.2 AU, until Jupiter's orbit. Here the two families of Trojan asteroids can be found, which, at least for objects larger than 1 km, are approximately as numerous as the asteroids of the asteroid belt.[70]

New families

Some asteroid families have formed recently, in astronomical terms. The Karin Cluster apparently formed about 5.7 million years ago from a collision with a 33 km radius progenitor asteroid.[71] The Veritas family formed about 8.3 million years ago; evidence includes interplanetary dust recovered from ocean sediment.[72]

More recently, the Datura cluster appears to have formed about 450 thousand years ago from a collision with a main-belt asteroid. The age estimate is based on the probability of the members having their current orbits, rather than from any physical evidence. However, this cluster may have been a source for some zodiacal dust material.[73] Other recent cluster formations, such as the Iannini cluster (*circa* 1–5 million years ago), may have provided additional sources of this asteroid dust.[74]

Exploration

Artist's concept of the Dawn spacecraft with Vesta (left) and Ceres (right)

The first spacecraft to traverse the asteroid belt was Pioneer 10, which entered the region on July 16, 1972. At the time there was some concern that the debris in the belt would pose a hazard to the spacecraft, but it has since been safely traversed by 11 Earth-based craft without incident. Pioneer 11, Voyagers 1 and 2 and Ulysses passed through the belt without imaging any asteroids. Galileo imaged the asteroid 951 Gaspra in 1991 and 243 Ida in 1993, NEAR imaged 253 Mathilde in 1997, Cassini imaged 2685 Masursky in 2000, Stardust imaged 5535 Annefrank in 2002, New Horizons imaged 132524 APL in 2006, Rosetta imaged 2867 Šteins in 2008, and Dawn has been orbiting Vesta since July 2011.[75] Due to the low density of materials within the belt, the odds of a probe running into an asteroid are now estimated at less than one in a billion.[76]

Most belt asteroids imaged to date have come from brief flyby opportunities by probes headed for other targets. Only the Dawn, NEAR and Hayabusa missions have studied asteroids for a protracted period in orbit and at the surface. Dawn explored Vesta from July 2011 to September 2012, and is currently en route to Ceres for a 2015 rendezvous. If the probe is still operational after examining Ceres, an extended mission could allow additional exploration, possibly a flyby of Pallas.[77]

See also

- Asteroid mining
- Asteroids in astrology
- Asteroids in fiction
- Colonization of the asteroids
- Debris disk
- Kuiper belt

References

[1] Krasinsky, G. A.; Pitjeva, E. V.; Vasilyev, M. V.; Yagudina, and E. I. (July 2002). "Hidden Mass in the Asteroid Belt". *Icarus* **158** (1): 98–105. Bibcode 2002Icar..158...98K. doi:10.1006/icar.2002.6837.

[2] Pitjeva, E. V. (2005). "High-Precision Ephemerides of Planets—EPM and Determination of Some Astronomical Constants" (http://iau-comm4.jpl.nasa.gov/EPM2004.pdf) (PDF). *Solar System Research* **39** (3): 176. Bibcode 2005SoSyR..39..176P. doi:10.1007/s11208-005-0033-2. .

[3] For recent estimates of the masses of Ceres, 4 Vesta, 2 Pallas and 10 Hygiea, see the references in the infoboxes of their respective articles.

[4] Yeomans, Donald K. (July 13, 2006). "JPL Small-Body Database Browser" (http://ssd.jpl.nasa.gov/sbdb.cgi). NASA JPL. Archived (http://web.archive.org/web/20100929043420/http://ssd.jpl.nasa.gov/sbdb.cgi) from the original on 29 September 2010. . Retrieved 2010-09-27.

[5] Hilton, J. (2001). "When Did the Asteroids Become Minor Planets?" (http://www.usno.navy.mil/USNO/astronomical-applications/astronomical-information-center/minor-planets). *US Naval Observatory (USNO)*. . Retrieved 2007-10-01.

[6] "Dawn: A Journey to the Beginning of the Solar System" (http://www-ssc.igpp.ucla.edu/dawn/background.html). *Space Physics Center: UCLA*. 2005. . Retrieved 2007-11-03.

[7] Hoskin, Michael. "Bode's Law and the Discovery of Ceres" (http://www.astropa.unipa.it/HISTORY/hoskin.html). *Churchill College, Cambridge*. . Retrieved 2010-07-12.

[8] Linda T. Elkins-Tanton, *Asteroids, Meteorites, and Comets*, 2010:10

[9] "Call the police! The story behind the discovery of the asteroids". *Astronomy Now* (June 2007): 60–61.

[10] Pogge, Richard (2006). "An Introduction to Solar System Astronomy: Lecture 45: Is Pluto a Planet?" (http://www.astronomy.ohio-state.edu/~pogge/Ast161/Unit6/dwarfs.html). *An Introduction to Solar System Astronomy*. Ohio State University. . Retrieved 2007-11-11.

[11] Harper, Douglas (2010). "Asteroid" (http://www.etymonline.com/index.php?search=asteroid&searchmode=none). *Online Etymology Dictionary*. Etymology Online. . Retrieved 2011-04-15.

[12] DeForest, Jessica (2000). "Greek and Latin Roots" (http://www.msu.edu/~defores1/gre/roots/gre_rts_afx2.htm). Michigan State University. Archived (http://web.archive.org/web/20070812155525/http://www.msu.edu/~defores1/gre/roots/gre_rts_afx2.htm) from the original on 12 August 2007. . Retrieved 2007-07-25.

[13] Cunningham, Clifford (1984). "William Hershel and the First Two Asteroids". *Dance Hall Observatory, Ontario* **11**: 3. Bibcode 1984MPBu...11....3C.

[14] Staff (2002). "Astronomical Serendipity" (http://dawn.jpl.nasa.gov/DawnCommunity/flashbacks/fb_06.asp). NASA JPL. . Retrieved 2007-04-20.

[15] "Is it a coincidence that most of the planets fall within the Titius-Bode law's boundaries?" (http://www.astronomy.com/asy/default.aspx?c=a&id=4494). *astronomy.com*. . Retrieved 2007-10-16.

[16] von Humboldt, Alexander (1850). *Cosmos: A Sketch of a Physical Description of the Universe*. **1**. Harper & Brothers, New York (NY). p. 44. ISBN 0-8018-5503-9.

[17] Mann, Robert James (1852). *A Guide to the Knowledge of the Heavens*. Jarrold. p. 171. and 1853, p. 216

[18] "Further Investigation relative to the form, the magnitude, the mass, and the orbit of the Asteroid Planets" (http://books.google.com/?id=hhQAAAAAMAAJ&pg=PA191&dq=asteroid+belt). *The Edinburgh New Philosophical Journal* **5**: 191. January–April 1857. .: "[Professor Peirce] then observed that the analogy between the ring of Saturn and the belt of the asteroids was worthy of notice."

[19] Hughes, David W. (2007). "A Brief History of Asteroid Spotting" (http://www.open2.net/sciencetechnologynature/planetsbeyond/asteroids/history.html). BBC. . Retrieved 2007-04-20.

[20] Moore, Patrick; Rees, Robin (2011). *Patrick Moore's Data Book of Astronomy* (2nd ed.). Cambridge University Press. p. 156. ISBN 0-521-89935-4.

[21] Manley, Scott (August 25, 2010). *Asteroid Discovery from 1980 to 2010* (http://www.youtube.com/watch?v=S_d-gs0WoUw). . Retrieved 2011-04-15.

[22] "MPC Archive Statistics" (http://www.minorplanetcenter.org/iau/lists/ArchiveStatistics.html). IAU Minor Planet Center. . Retrieved 2011-04-04.

[23] "A Brief History of Asteroid Spotting" (http://www.open2.net/sciencetechnologynature/planetsbeyond/asteroids/history.html). *Open2.net*. . Retrieved 2007-05-15.

[24] Masetti, M.; and Mukai, K. (December 1, 2005). "Origin of the Asteroid Belt" (http://imagine.gsfc.nasa.gov/docs/ask_astro/answers/980810a.html). NASA Goddard Spaceflight Center. . Retrieved 2007-04-25.

[25] Watanabe, Susan (July 20, 2001). "Mysteries of the Solar Nebula" (http://www.jpl.nasa.gov/news/features.cfm?feature=520). NASA. . Retrieved 2007-04-02.

[26] Petit, J.-M.; Morbidelli, A.; and Chambers, J. (2001). "The Primordial Excitation and Clearing of the Asteroid Belt" (http://www.gps.caltech.edu/classes/ge133/reading/asteroids.pdf) (PDF). *Icarus* **153** (2): 338–347. Bibcode 2001Icar..153..338P. doi:10.1006/icar.2001.6702. Archived (http://web.archive.org/web/20070221085835/http://www.gps.caltech.edu/classes/ge133/reading/asteroids.pdf) from the original on 21 February 2007. . Retrieved 2007-03-22.

[27] Edgar, R.; and Artymowicz, P. (2004). "Pumping of a Planetesimal Disc by a Rapidly Migrating Planet" (http://www.astro.su.se/~pawel/edgar+artymowicz.pdf) (PDF). *Monthly Notices of the Royal Astronomical Society* **354** (3): 769–772. arXiv:astro-ph/0409017. Bibcode 2004MNRAS.354..769E. doi:10.1111/j.1365-2966.2004.08238.x. . Retrieved 2007-04-16.

[28] Scott, E. R. D. (March 13–17, 2006). "Constraints on Jupiter's Age and Formation Mechanism and the Nebula Lifetime from Chondrites and Asteroids" (http://adsabs.harvard.edu/abs/2006LPI....37.2367S). *Proceedings 37th Annual Lunar and Planetary Science Conference*. League City, Texas: Lunar and Planetary Society. . Retrieved 2007-04-16.

[29] Taylor, G. J.; Keil, K.; McCoy, T.; Haack, H.; and Scott, E. R. D.; Keil; McCoy; Haack; Scott (1993). "Asteroid differentiation – Pyroclastic volcanism to magma oceans". *Meteoritics* **28** (1): 34–52. Bibcode 1993Metic..28...34T.

[30] Kelly, Karen (2007). "U of T researchers discover clues to early solar system" (http://webapps.utsc.utoronto.ca/ose/story.php?id=665). *University of Toronto*. . Retrieved 2010-07-12.

[31] Clark, B. E.; Hapke, B.; Pieters, C.; Britt, D.; Hapke; Pieters; Britt (2002). "Asteroid Space Weathering and Regolith Evolution". *University of Arizona*: 585. Bibcode 2002aste.conf..585C. Gaffey, Michael J. (1996). "The Spectral and Physical Properties of Metal in Meteorite Assemblages: Implications for Asteroid Surface Materials". *Icarus (ISSN 0019-1035)* **66** (3): 468. Bibcode 1986Icar...66..468G. doi:10.1016/0019-1035(86)90086-2. Keil, K. (2000). "Thermal alteration of asteroids: evidence from meteorites" (http://www.ingentaconnect.com/content/els/00320633/2000/00000048/00000010/art00054). *Planetary and Space Science*. . Retrieved 2007-11-08. Baragiola, R. A.; Duke, C. A.; Loeffler, M.; McFadden, L. A.; and Sheffield, J.; Duke; Loeffler; McFadden; Sheffield (2003). "Impact of ions and micrometeorites on mineral surfaces: Reflectance changes and production of atmospheric species in airless solar system bodies". *EGS - AGU - EUG Joint Assembly*: 7709. Bibcode 2003EAEJA.....7709B.

[32] Chapman, C. R.; Williams, J. G.; Hartmann, W. K. (1978). "The asteroids". *Annual review of astronomy and astrophysics* **16**: 33–75. Bibcode 1978ARA&A..16...33C. doi:10.1146/annurev.aa.16.090178.000341.

[33] Kracher, A. (2005). "Asteroid 433 Eros and partially differentiated planetesimals: bulk depletion versus surface depletion of sulfur" (http://www.cosis.net/abstracts/EGU05/03788/EGU05-J-03788.pdf) (PDF). *Ames Laboratory*. Archived (http://web.archive.org/web/20071128200221/http://www.cosis.net/abstracts/EGU05/03788/EGU05-J-03788.pdf) from the original on 28 November 2007. . Retrieved 2007-11-08.

[34] Stiles, Lori (September 15, 2005). "Asteroids Caused the Early Inner Solar System Cataclysm" (http://uanews.org/cgi-bin/WebObjects/UANews.woa/7/wa/SRStoryDetails?ArticleID=11692). University of Arizona News. . Retrieved 2007-04-18.

[35] Alfvén, H.; Arrhenius, G. (1976). "The Small Bodies" (http://history.nasa.gov/SP-345/ch4.htm). *SP-345 Evolution of the Solar System*. NASA. Archived (http://web.archive.org/web/20070513081833/http://history.nasa.gov/SP-345/ch4.htm) from the original on 13 May 2007. . Retrieved 2007-04-12.

[36] Spratt, Christopher E. (April 1990). "The Hungaria group of minor planets". *Journal of the Royal Astronomical Society of Canada* **84**: 123–131. Bibcode 1990JRASC..84..123S.

[37] Lecar, M.; Podolak, M.; Sasselov, D.; Chiang, E. (2006). "Infrared cirrus – New components of the extended infrared emission". *The Astrophysical Journal* **640** (2): 1115–1118. arXiv:astro-ph/0602217. Bibcode 2006ApJ...640.1115L. doi:10.1086/500287.

[38] Berardelli, Phil (March 23, 2006). "Main-Belt Comets May Have Been Source Of Earths Water" (http://www.spacedaily.com/reports/Main_Belt_Comets_May_Have_Been_Source_Of_Earths_Water.html). Space Daily. Archived (http://web.archive.org/web/20071011171815/http://spacedaily.com/reports/Main_Belt_Comets_May_Have_Been_Source_Of_Earths_Water.html) from the original on 11 October 2007. . Retrieved 2007-10-27.

[39] Lakdawalla, Emily (April 28, 2006). "Discovery of a Whole New Type of Comet" (http://www.planetary.org/blog/article/00000551/). The Planetary Society. Archived (http://web.archive.org/web/20070501211319/http://www.planetary.org/blog/article/00000551/) from the original on 1 May 2007. . Retrieved 2007-04-20.

[40] Yeomans, Donald K. (April 26, 2007). "JPL Small-Body Database Search Engine" (http://ssd.jpl.nasa.gov/sbdb_query.cgi). NASA JPL. . Retrieved 2007-04-26. – search for asteroids in the main belt regions with a diameter >100.

[41] Tedesco, E. F.; and Desert, F.-X. (2002). "The Infrared Space Observatory Deep Asteroid Search". *The Astronomical Journal* **123** (4): 2070–2082. Bibcode 2002AJ....123.2070T. doi:10.1086/339482.

[42] Williams, Gareth (September 25, 2010). "Distribution of the Minor Planets" (http://www.minorplanetcenter.org/iau/lists/MPDistribution.html). Minor Planets Center. . Retrieved 2010-10-27.

[43] Wiegert, P.; Balam, D.; Moss, A.; Veillet, C.; Connors, M.; and Shelton, I. (2007). "Evidence for a Color Dependence in the Size Distribution of Main-Belt Asteroids" (http://astro.uwo.ca/~wiegert/papers/2007AJ.133.1609.pdf). *The Astronomical Journal* **133** (4): 1609–1614. arXiv:astro-ph/0611310. Bibcode 2007AJ....133.1609W. doi:10.1086/512128. . Retrieved 2008-09-06.

[44] Clark, B. E. (1996). "New News and the Competing Views of Asteroid Belt Geology". *Lunar and Planetary Science* **27**: 225–226. Bibcode 1996LPI....27..225C.

[45] Margot, J. L.; and Brown, M. E. (2003). "A Low-Density M-type Asteroid in the Main Belt". *Science* **300** (5627): 1939–1942. Bibcode 2003Sci...300.1939M. doi:10.1126/science.1085844. PMID 12817147.

[46] Lang, Kenneth R. (2003). "Asteroids and meteorites" (http://ase.tufts.edu/cosmos/print_images.asp?id=15). NASA's Cosmos. . Retrieved 2007-04-02.

[47] Mueller, M.; Harris, A. W.; Delbo, M.; and the MIRSI Team; Harris; Delbo (2005). "21 Lutetia and other M-types: Their sizes, albedos, and thermal properties from new IRTF measurements". *Bulletin of the American Astronomical Society* **37**: 627. Bibcode 2005DPS....37.0702M.

[48] Duffard, R. D.; Roig, F. (July 14–18, 2008). "Two New Basaltic Asteroids in the Main Belt?". *Asteroids, Comets, Meteors 2008*. Baltimore, Maryland. arXiv:0704.0230. Bibcode 2008LPICo1405.8154D.

[49] Than, Ker (2007). "Strange Asteroids Baffle Scientists" (http://www.space.com/scienceastronomy/070821_basalt_asteroid.html). *space.com*. . Retrieved 2007-10-14.

[50] Low, F. J.; *et al.* (1984). "Infrared cirrus – New components of the extended infrared emission". *Astrophysical Journal, Part 2 – Letters to the Editor* **278**: L19–L22. Bibcode 1984ApJ...278L..19L. doi:10.1086/184213.

[51] "Interview with David Jewitt" (http://www.youtube.com/watch?v=B1W4NTmI5Bk). Youtube.com. 2007-01-05. . Retrieved 2011-05-21.

[52] This value was obtained by a simple count up of all bodies in that region using data for 120437 numbered minor planets from the Minor Planet Center orbit database (http://www.minorplanetcenter.org/iau/MPCORB.html), dated February 8, 2006.

[53] Fernie, J. Donald (1999). "The American Kepler" (http://www.americanscientist.org/issues/pub/1999/9/the-american-kepler/2). *The Americal Scientist* **87** (5): 398. . Retrieved 2007-02-04.

[54] Liou, Jer-Chyi; and Malhotra, Renu (1997). "Depletion of the Outer Asteroid Belt" (http://www.sciencemag.org/cgi/content/full/275/5298/375). *Science* **275** (5298): 375–377. Bibcode 1997Sci...275..375L. doi:10.1126/science.275.5298.375. PMID 8994031. . Retrieved 2007-08-01.

[55] McBride, N.; and Hughes, D. W.; Hughes (1990). "The spatial density of asteroids and its variation with asteroidal mass". *Monthly Notices of the Royal Astronomical Society* **244**: 513–520. Bibcode 1990MNRAS.244..513M.

[56] Ferraz-Mello, S. (June 14–18, 1993). "Kirkwood Gaps and Resonant Groups" (http://adsabs.harvard.edu/abs/1994IAUS..160..175F). *proceedings of the 160th International Astronomical Union*. Belgirate, Italy: Kluwer Academic Publishers. pp. 175–188. . Retrieved 2007-03-28.

[57] Klacka, Jozef (1992). "Mass distribution in the asteroid belt". *Earth, Moon, and Planets* **56** (1): 47–52. Bibcode 1992EM&P...56...47K. doi:10.1007/BF00054599.

[58] Backman, D. E. (March 6, 1998). "Fluctuations in the General Zodiacal Cloud Density" (http://astrobiology.arc.nasa.gov/workshops/zodiac/backman/backman_toc.html). *Backman Report*. NASA Ames Research Center. . Retrieved 2007-04-04.

[59] Reach, William T. (1992). "Zodiacal emission. III – Dust near the asteroid belt". *Astrophysical Journal* **392** (1): 289–299. Bibcode 1992ApJ...392..289R. doi:10.1086/171428.

[60] Kingsley, Danny (May 1, 2003). "Mysterious meteorite dust mismatch solved" (http://abc.net.au/science/news/stories/s843594.htm). ABC Science. . Retrieved 2007-04-04.

[61] "Meteors and Meteorites" (http://www.nasa.gov/pdf/145945main_Meteors.Meteorites.Lithograph.pdf). NASA. . Retrieved 2012-01-12.

[62] "Breakup event in the main asteroid belt likely caused dinosaur extinction 65 million years ago" (http://www.physorg.com/news108218928.html). *Southwest Research Institute*. 2007. Archived (http://web.archive.org/web/20071011151507/http://physorg.com/news108218928.html) from the original on 11 October 2007. . Retrieved 2007-10-14.

[63] Hughes, David W. (2007). "Finding Asteroids In Space" (http://www.open2.net/sciencetechnologynature/planetsbeyond/asteroids/finding.html). BBC. . Retrieved 2007-04-20.

[64] Lemaitre, Anne (August 31-September 4, 2004). "Asteroid family classification from very large catalogues" (http://adsabs.harvard.edu/abs/2005dpps.conf..135L). *Proceedings Dynamics of Populations of Planetary Systems*. Belgrade, Serbia and Montenegro: Cambridge University Press. pp. 135–144. . Retrieved 2007-04-15.

[65] Martel, Linda M. V. (March 9, 2004). "Tiny Traces of a Big Asteroid Breakup" (http://www.psrd.hawaii.edu/Mar04/fossilMeteorites.html). Planetary Science Research Discoveries. Archived (http://web.archive.org/web/20070401115157/http://www.psrd.hawaii.edu/Mar04/fossilMeteorites.html) from the original on 1 April 2007. . Retrieved 2007-04-02.

[66] Drake, Michael J. (2001). "The eucrite/Vesta story". *Meteoritics & Planetary Science* **36** (4): 501–513. Bibcode 2001M&PS...36..501D. doi:10.1111/j.1945-5100.2001.tb01892.x.

[67] Love, S. G.; and Brownlee, D. E. (1992). "The IRAS dust band contribution to the interplanetary dust complex – Evidence seen at 60 and 100 microns". *Astronomical Journal* **104** (6): 2236–2242. Bibcode 1992AJ....104.2236L. doi:10.1086/116399.

[68] Spratt, Christopher E. (1990). "The Hungaria group of minor planets". *Journal of the Royal Astronomical Society of Canada* **84** (2): 123–131. Bibcode 1990JRASC..84..123S.

[69] Carvano, J. M.; Lazzaro, D.; Mothé-Diniz, T.; Angeli, C. A.; and Florczak, M. (2001). "Spectroscopic Survey of the Hungaria and Phocaea Dynamical Groups". *Icarus* **149** (1): 173–189. Bibcode 2001Icar..149..173C. doi:10.1006/icar.2000.6512.

[70] Dymock, Roger (2010). *Asteroids and Dwarf Planets and How to Observe Them* (http://books.google.com/books?id=vQcAnwt_87sC&pg=PA24). Astronomers' Observing Guides. Springer. p. 24. ISBN 1-4419-6438-X. . Retrieved 2011-04-04.

[71] Nesvorný, David; *et al.* (August 2006). "Karin cluster formation by asteroid impact". *Icarus* **183** (2): 296–311. Bibcode 2006Icar..183..296N. doi:10.1016/j.icarus.2006.03.008.

[72] McKee, Maggie (January 18, 2006). "Eon of dust storms traced to asteroid smash" (http://space.newscientist.com/channel/solar-system/comets-asteroids/dn8603). New Scientist Space. . Retrieved 2007-04-15.

[73] Nesvorný, D.; Vokrouhlick, D.; and Bottke, W. F. (2006). "The Breakup of a Main-Belt Asteroid 450 Thousand Years Ago" (http://www.sciencemag.org/cgi/content/full/312/5779/1490). *Science* **312** (5779): 1490. Bibcode 2006Sci...312.1490N. doi:10.1126/science.1126175. PMID 16763141. . Retrieved 2007-04-15.

[74] Nesvorný, D.; Bottke, W. F.; Levison, H. F.; and Dones, L. (2003). "Recent Origin of the Solar System Dust Bands" (http://iopscience.iop.org/0004-637X/591/1/486/pdf/0004-637X_591_1_486.pdf). *The Astrophysical Journal* **591** (1): 486–497. Bibcode 2003ApJ...591..486N. doi:10.1086/374807. . Retrieved 2007-04-15.

[75] Barucci, M. A.; Fulchignoni, M.; and Rossi, A. (2007). "Rosetta Asteroid Targets: 2867 Steins and 21 Lutetia". *Space Science Reviews* **128** (1–4): 67–78. Bibcode 2007SSRv..128...67B. doi:10.1007/s11214-006-9029-6.

[76] Stern, Alan (June 2, 2006). "New Horizons Crosses The Asteroid Belt" (http://www.spacedaily.com/reports/New_Horizons_Crosses_The_Asteroid_Belt.html). Space Daily. . Retrieved 2007-04-14.
[77] Staff (April 10, 2007). "Dawn Mission Home Page" (http://dawn.jpl.nasa.gov/). NASA JPL. Archived (http://web.archive.org/web/20070411053350/http://dawn.jpl.nasa.gov/) from the original on 11 April 2007. . Retrieved 2007-04-14.

Further reading

- Elkins-Tanton, Linda T. (2006). *Asteroids, Meteorites, and Comets* (First ed.). New York: Chelsea House. ISBN 0-8160-5195-X.

External links

- Asteroid Discovery from 1980 to 2010 (http://www.youtube.com/watch?v=S_d-gs0WoUw)
- Arnett, William A. (February 26, 2006). "Asteroids" (http://www.nineplanets.org/asteroids.html). The Nine Planets. Archived (http://web.archive.org/web/20070418220152/http://www.nineplanets.org/asteroids.html) from the original on 18 April 2007. Retrieved 2007-04-20.
- Asteroids Page (http://solarsystem.nasa.gov/planets/profile.cfm?Object=Asteroids) at NASA's Solar System Exploration (http://solarsystem.nasa.gov)
- Cain, Fraser. "The Asteroid Belt" (http://www.astronomycast.com/astronomy/episode-55-the-asteroid-belt/). Universe Today. Archived (http://web.archive.org/web/20080307033324/http://www.astronomycast.com/astronomy/episode-55-the-asteroid-belt/) from the original on 7 March 2008. Retrieved 2008-04-01.
- Hsieh, Henry H. (March 1, 2006). "Main-Belt Comets" (http://star.pst.qub.ac.uk/~hhh/mbcs.shtml). University of Hawaii. Retrieved 2007-04-20.
- "Main Asteroid Belt" (http://www.solstation.com/stars/asteroid.htm). Sol Company. Archived (http://web.archive.org/web/20070515163455/http://www.solstation.com/stars/asteroid.htm) from the original on 15 May 2007. Retrieved 2007-04-20.
- Munsell, Kirk (September 16, 2005). "Asteroids: Overview" (http://solarsystem.nasa.gov/planets/profile.cfm?Object=Asteroids). NASA's Solar System Exploration (http://solarsystem.nasa.gov). Archived (http://web.archive.org/web/20070524185042/http://solarsystem.nasa.gov/planets/profile.cfm?Object=Asteroids) from the original on 24 May 2007. Retrieved 2007-05-26.
- Plots of eccentricity vs. semi-major axis (http://burro.astr.cwru.edu/stu/media/asteroid_all_axisvecc.jpg) and inclination vs. semi-major axis (http://burro.astr.cwru.edu/stu/media/asteroid_all_axisvincl.jpg) at Asteroid Dynamic Site
- Staff (October 31, 2006). "Asteroids" (http://nssdc.gsfc.nasa.gov/planetary/planets/asteroidpage.html). NASA. Archived (http://web.archive.org/web/20070411201013/http://nssdc.gsfc.nasa.gov/planetary/planets/asteroidpage.html) from the original on 11 April 2007. Retrieved 2007-04-20.
- Staff (2007). "Space Topics: Asteroids and Comets" (http://www.planetary.org/explore/topics/asteroids_and_comets/facts.html). The Planetary Society. Archived (http://web.archive.org/web/20070428143402/http://www.planetary.org/explore/topics/asteroids_and_comets/facts.html) from the original on 28 April 2007. Retrieved 2007-04-20.

Asteroid

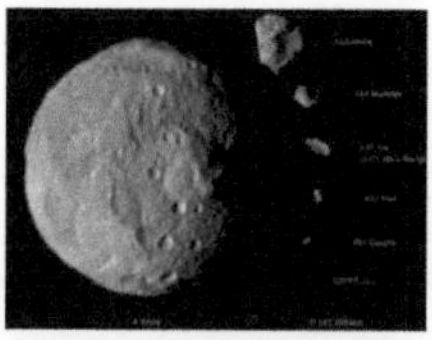

A composite image, to scale, of the asteroids which have been imaged at high resolution. As of 2011 they are, from largest to smallest: 4 Vesta, 21 Lutetia, 253 Mathilde, 243 Ida and its moon Dactyl, 433 Eros, 951 Gaspra, 2867 Šteins, 25143 Itokawa.

The largest asteroid in the previous image, Vesta (left), with Ceres (center) and Earth's Moon (right) shown to scale.

Asteroids are a class of small Solar System bodies in orbit around the Sun. They have also been called **planetoids**, especially the larger ones. These terms have historically been applied to any astronomical object orbiting the Sun that did not show the disk of a planet and was not observed to have the characteristics of an active comet, but as small objects in the outer Solar System were discovered, their volatile-based surfaces were found to more closely resemble comets, and so were often distinguished from traditional asteroids.[1] Thus the term *asteroid* has come increasingly to refer specifically to the small bodies of the inner Solar System out to the orbit of Jupiter, which are usually rocky or metallic. They are grouped with the outer bodies—centaurs, Neptune trojans, and trans-Neptunian objects—as minor planets, which is the term preferred in astronomical circles.[2] This article will restrict the use of the term "asteroid" to the minor planets of the inner Solar System.

There are millions of asteroids, many thought to be the shattered remnants of planetesimals, bodies within the young Sun's solar nebula that never grew large enough to become planets.[3] A large majority of known asteroids orbit in the asteroid belt between the orbits of Mars and Jupiter or co-orbital with Jupiter (the Jupiter Trojans). However, other orbital families exist with significant populations, including the near-Earth asteroids. Individual asteroids are classified by their characteristic spectra, with the majority falling into three main groups: C-type, S-type, and M-type. These were named after and are generally identified with carbon-rich, stony, and metallic compositions, respectively.

Naming

A newly discovered asteroid is given a provisional designation (such as 2002 AT_4) consisting of the year of discovery and an alphanumeric code indicating the half-month of discovery and the sequence within that half-month. Once an asteroid's orbit has been confirmed, it is given a number, and later may also be given a name (e.g. 433 Eros). The formal naming convention uses parentheses around the number (e.g. (433) Eros), but dropping the parentheses is quite common. Informally, it is common to drop the number altogether, or to drop it after the first mention when a name is repeated in running text.

Symbols

The first asteroids to be discovered were assigned iconic symbols like the ones traditionally used to designate the planets. By 1855 there were two dozen asteroid symbols, which often occurred in multiple variants.[4]

Asteroid	Symbol		Year
Ceres		Ceres' scythe, reversed to double as the letter *C*	1801
2 Pallas		Athena's (Pallas') spear	1801
3 Juno		A star mounted on a scepter, for Juno, the Queen of Heaven	1804
4 Vesta		The altar and sacred fire of Vesta	1807
5 Astraea		A scale, or an inverted anchor, symbols of justice	1845
6 Hebe		Hebe's cup	1847
7 Iris		A rainbow (*iris*) and a star	1847
8 Flora		A flower (*flora*) (spec. the Rose of England)	1847
9 Metis		The eye of wisdom and a star	1848
10 Hygiea		Hygiea's serpent and a star, or the Rod of Asclepius	1849
11 Parthenope		A harp, or a fish and a star; symbols of the sirens	1850
12 Victoria		The laurels of victory and a star	1850
13 Egeria		A shield, symbol of Egeria's protection, and a star	1850
14 Irene		A dove carrying an olive-branch (symbol of *irene* 'peace') with a star on its head,[5] or an olive branch, a flag of truce, and a star	1851
15 Eunomia		A heart, symbol of good order (*eunomia*), and a star	1851
16 Psyche		A butterfly's wing, symbol of the soul (*psyche*), and a star	1852
17 Thetis		A dolphin, symbol of Thetis, and a star	1852
18 Melpomene		The dagger of Melpomene, and a star	1852
19 Fortuna		The wheel of fortune and a star	1852
26 Proserpina		Proserpina's pomegranate	1853
28 Bellona		Bellona's whip and lance[6]	1854
29 Amphitrite		The shell of Amphitrite and a star	1854
35 Leukothea		A lighthouse beacon, symbol of Leucothea[7]	1855
37 Fides	†	The cross of faith (*fides*)[8]	1855

In 1851,[9] after the fifteenth asteroid (Eunomia) had been discovered, Johann Franz Encke made a major change in the upcoming 1854 edition of the *Berliner Astronomisches Jahrbuch* (BAJ, *Berlin Astronomical Yearbook*). He introduced a disk (circle), a traditional symbol for a star, as the generic symbol for an asteroid. The circle was then numbered in order of discovery to indicate a specific asteroid (although he assigned ① to the fifth, Astraea, while continuing to designate the first four only with their existing iconic symbols). The numbered-circle convention was quickly adopted by the astronomical community, and the next asteroid to be discovered (16 Psyche, in 1852) was the first to be designated in this manner at the time of its discovery. However, Psyche was also given an iconic symbol, as were a few other asteroids discovered over the next few years (see chart above). 20 Massalia was the first asteroid that was not assigned a symbol, and no additional iconic symbols were created after the 1855 discovery of 37 Fides.[10] That year Astraea's number was bumped up to ⑤, but Ceres through Vesta would not be listed by their numbers until the 1867 edition. The circle would become a pair of parentheses, and the parentheses sometimes omitted altogether over the next few decades, leading to the modern convention.[5]

Discovery

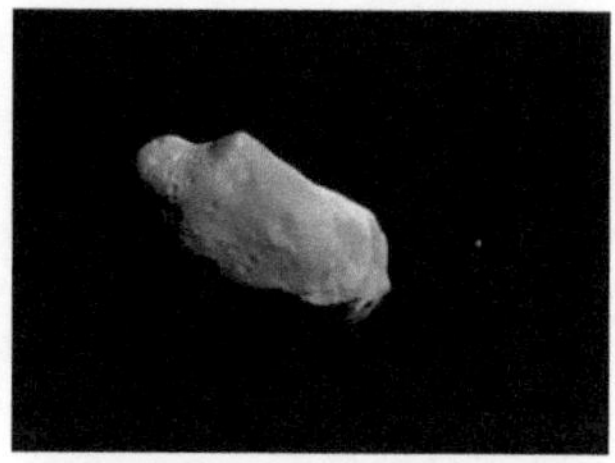

243 Ida and its moon Dactyl. Dactyl is the first satellite of an asteroid to be discovered.

The first asteroid to be discovered, Ceres, was found in 1801 by Giuseppe Piazzi, and was originally considered to be a new planet.[11] This was followed by the discovery of other similar bodies, which, with the equipment of the time, appeared to be points of light, like stars, showing little or no planetary disc, though readily distinguishable from stars due to their apparent motions. This prompted the astronomer Sir William Herschel to propose the term "asteroid", coined in Greek as αστεροειδής *asteroeidēs* 'star-like, star-shaped', from ancient Greek ἀστήρ *astēr* 'star, planet'. In the early second half of the nineteenth century, the terms "asteroid" and "planet" (not always qualified as "minor") were still used interchangeably; for example, the *Annual of Scientific Discovery for 1871* [12], page 316, reads "Professor J. Watson has been awarded by the Paris Academy of Sciences, the astronomical prize, Lalande foundation, for the discovery of eight new asteroids in one year. The planet Lydia (No. 110), discovered by M. Borelly at the Marseilles Observatory [...] M. Borelly had previously discovered two planets bearing the numbers 91 and 99 in the system of asteroids revolving between Mars and Jupiter".

Historical methods

Asteroid discovery methods have dramatically improved over the past two centuries.

In the last years of the 18th century, Baron Franz Xaver von Zach organized a group of 24 astronomers to search the sky for the missing planet predicted at about 2.8 AU from the Sun by the Titius-Bode law, partly because of the discovery, by Sir William Herschel in 1781, of the planet Uranus at the distance predicted by the law. This task required that hand-drawn sky charts be prepared for all stars in the zodiacal band down to an agreed-upon limit of faintness. On subsequent nights, the sky would be charted again and any moving object would, hopefully, be spotted. The expected motion of the missing planet was about 30 seconds of arc per hour, readily discernible by observers.

The first object, Ceres, was not discovered by a member of the group, but rather by accident in 1801 by Giuseppe Piazzi, director of the observatory of Palermo in Sicily. He discovered a new star-like object in Taurus and followed the displacement of this object during several nights. His colleague, Carl Friedrich Gauss, used these observations to find the exact distance from this unknown object to the Earth. Gauss' calculations placed the object between the planets Mars and Jupiter. Piazzi named it after Ceres, the Roman goddess of agriculture.

Three other asteroids (2 Pallas, 3 Juno, and 4 Vesta) were discovered over the next few years, with Vesta found in 1807. After eight more years of fruitless searches, most astronomers assumed that there were no more and abandoned any further searches.

However, Karl Ludwig Hencke persisted, and began searching for more asteroids in 1830. Fifteen years later, he found 5 Astraea, the first new asteroid in 38 years. He also found 6 Hebe less than two years later. After this, other astronomers joined in the search and at least one new asteroid was discovered every year after that (except the wartime year 1945). Notable asteroid hunters of this early era were J. R. Hind, Annibale de Gasparis, Robert Luther, H. M. S. Goldschmidt, Jean Chacornac, James Ferguson, Norman Robert Pogson, E. W. Tempel, J. C. Watson, C. H. F. Peters, A. Borrelly, J. Palisa, the Henry brothers and Auguste Charlois.

In 1891, however, Max Wolf pioneered the use of astrophotography to detect asteroids, which appeared as short streaks on long-exposure photographic plates. This dramatically increased the rate of detection compared with earlier visual methods: Wolf alone discovered 248 asteroids, beginning with 323 Brucia, whereas only slightly more than 300 had been discovered up to that point. It was known that there were many more, but most astronomers did not bother with them, calling them "vermin of the skies", a phrase due to Edmund Weiss.[13] Even a century later, only a few thousand asteroids were identified, numbered and named.

Manual methods of the 1900s and modern reporting

Until 1998, asteroids were discovered by a four-step process. First, a region of the sky was photographed by a wide-field telescope, or Astrograph. Pairs of photographs were taken, typically one hour apart. Multiple pairs could be taken over a series of days. Second, the two films or plates of the same region were viewed under a stereoscope. Any body in orbit around the Sun would move slightly between the pair of films. Under the stereoscope, the image of the body would seem to float slightly above the background of stars. Third, once a moving body was identified, its location would be measured precisely using a digitizing microscope. The location would be measured relative to known star locations.[14]

These first three steps do not constitute asteroid discovery: the observer has only found an apparition, which gets a provisional designation, made up of the year of discovery, a letter representing the half-month of discovery, and finally a letter and a number indicating the discovery's sequential number (example: 1998 FJ_{74}).

The last step of discovery is to send the locations and time of observations to the Minor Planet Center, where computer programs determine whether an apparition ties together earlier apparitions into a single orbit. If so, the object receives a catalogue number and the observer of the first apparition with a calculated orbit is declared the discoverer, and granted the honor of naming the object subject to the approval of the International Astronomical Union.

Computerized methods

2004 FH is the center dot being followed by the sequence; the object that flashes by during the clip is an artificial satellite.

There is increasing interest in identifying asteroids whose orbits cross Earth's, and that could, given enough time, collide with Earth (see Earth-crosser asteroids). The three most important groups of near-Earth asteroids are the Apollos, Amors, and Atens. Various asteroid deflection strategies have been proposed, as early as the 1960s.

The near-Earth asteroid 433 Eros had been discovered as long ago as 1898, and the 1930s brought a flurry of similar objects. In order of discovery, these were: 1221 Amor, 1862 Apollo, 2101 Adonis, and finally 69230 Hermes, which approached within 0.005 AU of the Earth in 1937. Astronomers began to realize the possibilities of Earth impact.

Two events in later decades increased the alarm: the increasing acceptance of Walter Alvarez' hypothesis that an impact event resulted in the Cretaceous–Paleogene extinction, and the 1994 observation of Comet Shoemaker-Levy 9 crashing into Jupiter. The U.S. military also declassified the information that its military satellites, built to detect nuclear explosions, had detected hundreds of upper-atmosphere impacts by objects ranging from one to 10 metres across.

All these considerations helped spur the launch of highly efficient automated systems that consist of Charge-Coupled Device (CCD) cameras and computers directly connected to telescopes. Since 1998, a large majority of the asteroids have been discovered by such automated systems. A list of teams using such automated systems includes:[15]

- The Lincoln Near-Earth Asteroid Research (LINEAR) team
- The Near-Earth Asteroid Tracking (NEAT) team
- Spacewatch
- The Lowell Observatory Near-Earth-Object Search (LONEOS) team
- The Catalina Sky Survey (CSS)
- The Campo Imperatore Near-Earth Objects Survey (CINEOS) team
- The Japanese Spaceguard Association
- The Asiago-DLR Asteroid Survey (ADAS)

The LINEAR system alone has discovered 121,346 asteroids, as of March, 2011.[16] Among all the automated systems, 4711 near-Earth asteroids have been discovered[17] including over 600 more than 1 km (0.6 mi) in diameter.

Terminology

Traditionally, small bodies orbiting the Sun were classified as asteroids, comets or meteoroids, with anything smaller than ten metres across being called a meteoroid.[18] The term "asteroid" is ill-defined. It never had a formal definition, with the broader term minor planet being preferred by the International Astronomical Union from 1853 on. In 2006, the term "small Solar System body" was introduced to cover both most minor planets and comets.[19] Other languages prefer "planetoid" (Greek for "planet-like"), and this term is occasionally used in English for the larger asteroids. The word "planetesimal" has a similar meaning, but refers specifically to the small building blocks of the planets that existed when the Solar System was forming. The term "planetule" was coined by the geologist William Daniel Conybeare to describe minor planets,[20] but is not in common use. The three largest objects in the asteroid belt, Ceres, 2 Pallas, and 4 Vesta, grew to the stage of protoplanets. Ceres has been classified as a dwarf planet, the only one in the inner Solar System.

When found, asteroids were seen as a class of objects distinct from comets, and there was no unified term for the two until "small Solar System body" was coined in 2006. The main difference between an asteroid and a comet is that a comet shows a coma due to sublimation of near surface ices by solar radiation. A few objects have ended up being dual-listed because they were first classified as minor planets but later showed evidence of cometary activity. Conversely, some (perhaps all) comets are eventually depleted of their surface volatile ices and become asteroids. A further distinction is that comets typically have more eccentric orbits than most asteroids; most "asteroids" with notably eccentric orbits are probably dormant or extinct comets.[21]

For almost two centuries, from the discovery of Ceres in 1801 until the discovery of the first centaur, 2060 Chiron, in 1977, all known asteroids spent most of their time at or within the orbit of Jupiter, though a few such as 944 Hidalgo ventured far beyond Jupiter for part of their orbit. When astronomers started finding more small bodies that permanently resided further out than Jupiter, now called centaurs, they numbered them among the traditional asteroids, though there was debate over whether they should be classified as asteroids or as a new type of object. Then, when the first trans-Neptunian object, 1992 QB1, was discovered in 1992, and especially when large numbers of similar objects started turning up, new terms were invented to sidestep the issue: Kuiper-belt object, trans-Neptunian object, scattered-disc object, and so on. These inhabit the cold outer reaches of the Solar System where ices remain solid and comet-like bodies are not expected to exhibit much cometary activity; if centaurs or trans-Neptunian objects were to venture close to the Sun, their volatile ices would sublimate, and traditional approaches would classify them as comets and not asteroids.

The innermost of these are the Kuiper-belt objects, called "objects" partly to avoid the need to classify them as asteroids or comets.[22] They are believed to be predominantly comet-like in composition, though some may be more akin to asteroids.[23] Furthermore, most do not have the highly eccentric orbits associated with comets, and the ones so far discovered are larger than traditional comet nuclei. (The much more distant Oort cloud is hypothesized to be the main reservoir of dormant comets.) Other recent observations, such as the analysis of the cometary dust collected by the Stardust probe, are increasingly blurring the distinction between comets and asteroids,[24] suggesting "a continuum between asteroids and comets" rather than a sharp dividing line.[25]

The minor planets beyond Jupiter's orbit are sometimes also called "asteroids", especially in popular presentations.[26] However, it is becoming increasingly common for the term "asteroid" to be restricted to minor planets of the inner Solar System.[22] Therefore, this article will restrict itself for the most part to the classical asteroids: objects of the asteroid belt, Jupiter trojans, and near-Earth objects.

When the IAU introduced the class small Solar System bodies in 2006 to include most objects previously classified as minor planets and comets, they created the class of dwarf planets for the largest minor planets—those that have enough mass to have become ellipsoidal under their own gravity. According to the IAU, "the term 'minor planet' may still be used, but generally the term 'Small Solar System Body' will be preferred."[27] Currently only the largest object in the asteroid belt, Ceres, at about 950 km (590 mi) across, has been placed in the dwarf planet category, although there are several large asteroids (Vesta, Pallas, and Hygiea) that may be classified as dwarf planets when their shapes are better known.[28]

Formation

It is believed that planetesimals in the asteroid belt evolved much like the rest of the solar nebula until Jupiter neared its current mass, at which point excitation from orbital resonances with Jupiter ejected over 99% of planetesimals in the belt. Simulations and a discontinuity in spin rate and spectral properties suggest that asteroids larger than approximately 120 km (75 mi) in diameter accreted during that early era, whereas smaller bodies are fragments from collisions between asteroids during or after the Jovian disruption.[29] Ceres and Vesta grew large enough to melt and differentiate, with heavy metallic elements sinking to the core, leaving rocky minerals in the crust.[30]

In the Nice model, many Kuiper-belt objects are captured in the outer asteroid belt, at distances greater than 2.6 AU. Most were later ejected by Jupiter, but those that remained may be the D-type asteroids, and possibly include Ceres.[31]

Distribution within the Solar System

Various dynamical groups of asteroids have been discovered orbiting in the inner Solar System. Their orbits are perturbed by the gravity of other bodies in the Solar System and by the Yarkovsky effect. Significant populations include:

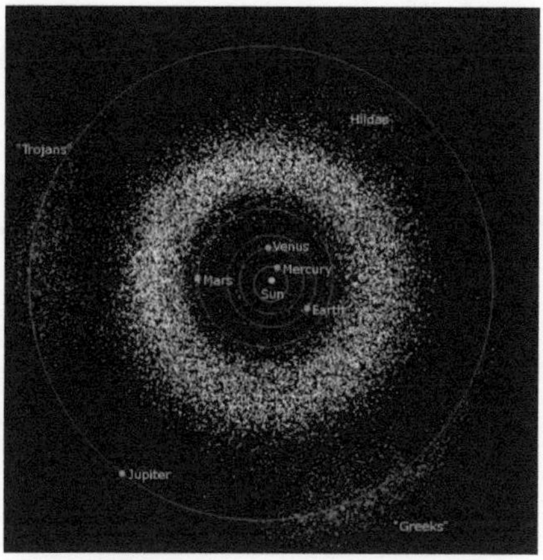

The asteroid belt (white) and the Trojan asteroids (green)

Asteroid belt

The majority of known asteroids orbit within the asteroid belt between the orbits of Mars and Jupiter, generally in relatively low-eccentricity (i.e., not very elongated) orbits. This belt is now estimated to contain between 1.1 and 1.9 million asteroids larger than 1 km (0.6 mi) in diameter,[32] and millions of smaller ones.[33] These asteroids may be remnants of the protoplanetary disk, and in this region the accretion of planetesimals into planets during the formative period of the Solar System was prevented by large gravitational perturbations by Jupiter.

Trojans

Trojan asteroids are a population that share an orbit with a larger planet or moon, but do not collide with it because they orbit in one of the two Lagrangian points of stability, L4 and L5, which lie 60° ahead of and behind the larger body.

The most significant population of Trojan asteroids are the Jupiter Trojans. Although fewer Jupiter Trojans have been discovered as of 2010, it is thought that they are as numerous as the asteroids in the asteroid belt.

A couple of trojans have also been found orbiting with Mars.[34]

Near-Earth asteroids

Near-Earth asteroids, or NEAs, are asteroids that have orbits that pass close to that of Earth. Asteroids that actually cross the Earth's orbital path are known as *Earth-crossers*. As of May 2010, 7,075 near-Earth asteroids are known and the number over one kilometre in diameter is estimated to be 500–1,000.

Characteristics

Size distribution

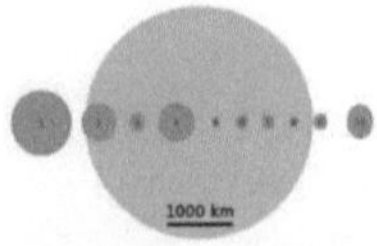

Sizes of the first ten asteroids to be discovered, compared to the Earth's Moon

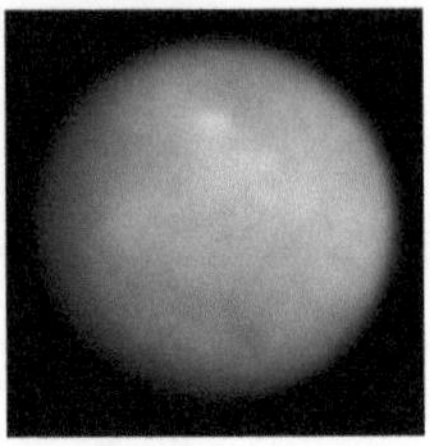

HST image of the dwarf planet Ceres

Asteroids vary greatly in size, from almost 1000 kilometres for the largest down to rocks just tens of metres across.[35] The three largest are very much like miniature planets: they are roughly spherical, have at least partly differentiated interiors,[36] and are thought to be surviving protoplanets. The vast majority, however, are much smaller and are irregularly shaped; they are thought to be either surviving planetesimals or fragments of larger bodies.

The dwarf planet Ceres is by far the largest asteroid, with a diameter of 975 km (610 mi). The next largest are 2 Pallas and 4 Vesta, both with diameters of just over 500 km (300 mi). Vesta is the only main-belt asteroid that can, on occasion, be visible to the naked eye. On some rare occasions, a near-Earth asteroid may briefly become visible without technical aid; see 99942 Apophis.

The mass of all the objects of the asteroid belt, lying between the orbits of Mars and Jupiter, is estimated to be about $2.8–3.2\times10^{21}$ kg, or about 4 percent of the mass of the Moon. Of this, Ceres comprises 0.95×10^{21} kg, a third of the total.[37] Adding in the next three most massive objects, Vesta (9%), Pallas (7%), and Hygiea (3%), brings this figure up to 51%; while the three after that, 511 Davida (1.2%), 704 Interamnia (1.0%), and 52 Europa (0.9%), only add another 3% to the total mass. The number of asteroids then increases rapidly as their individual masses decrease.

The number of asteroids decreases markedly with size. Although this generally follows a power law, there are 'bumps' at 5 km and 100 km, where more asteroids than expected from a logarithmic distribution are found.[38]

Approximate number of asteroids N larger than diameter D

D	100 m	300 m	500 m	1 km	3 km	5 km	10 km	30 km	50 km	100 km	200 km	300 km	500 km	900 km
N	~25,000,000	4,000,000	2,000,000	750,000	200,000	90,000	10,000	1,100	600	200	30	5	3	1

Largest asteroids

Although their location in the asteroid belt excludes them from planet status, the four largest objects, Ceres, Vesta, Pallas, and Hygiea, are remnant protoplanets that share many characteristics common to planets, and are atypical compared to the majority of "potato"-shaped asteroids.

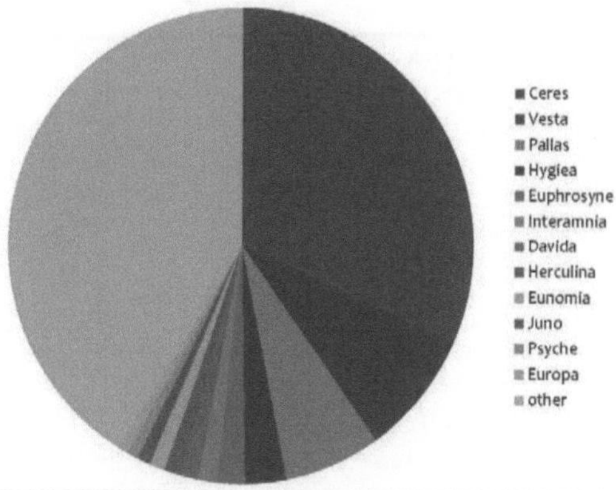

The relative masses of the twelve List of notable asteroids#Largest by masslargest asteroids known, "Recent Asteroid Mass Determinations". Maintained by Jim Baer. Last updated 2010-12-12. Access date 2011-09-02. The values of Juno and Herculina may be off by as much as 16%, and Euphrosyne by a third. The order of the lower eight may change as better data is acquired, but the values do not overlap with any known asteroid outside these twelve. compared to the remaining mass of the asteroid belt. Elena V. PitjevaPitjeva, E. V. (2005). "High-Precision Ephemerides of Planets—EPM and Determination of Some Astronomical Constants" (PDF). Solar System Research 39 (3): 184. Bibcode 2005SoSyR..39..176P. doi:10.1007/s11208-005-0033-2. .

Attributes of protoplanetary asteroids

Name	Orbital radius (AU)	Orbital period (years)	Inclination to ecliptic	Orbital eccentricity	Diameter (km)	Diameter (% of Moon)	Mass ($\times 10^{18}$ kg)	Mass (% of Ceres)	Density[41] g/cm^3	Rotation period (hr)	Axial tilt	Surface temperature
Vesta	2.36	3.63	7.1°	0.089	573×557×446 (mean 525)	15%	260	28%	3.44 ± 0.12	5.34	29°	85–270 K
Ceres	2.77	4.60	10.6°	0.079	975×975×909 (mean 952)	28%	940	100%	2.12 ± 0.04	9.07	≈ 3°	167 K
Pallas	2.77	4.62	34.8°	0.231	580×555×500 (mean 545)	16%	210	22%	2.71 ± 0.11	7.81	≈ 80°	164 K
Hygiea	3.14	5.56	3.8°	0.117	530×407×370 (mean 430)	12%	87	9%	2.76 ± 1.2	27.6	≈ 60°	164 K

Ceres is the only asteroid large enough for its gravity to force it into a spheroidal shape, and so, according to the IAU's 2006 resolution on the definition of a planet, it has been classified as a dwarf planet.[42] Vesta may eventually be so classified as well. Ceres has a much higher absolute magnitude than the other asteroids, of around 3.32,[43] and may possess a surface layer of ice.[44] Like the planets, Ceres is differentiated: it has a crust, a mantle and a core.[44] Vesta, too, has a differentiated interior, though it formed inside the Solar System's frost line, and so is devoid of water;[45] its composition is mainly of basaltic rock such as olivine.[46] Pallas is unusual in that, like Uranus, it rotates on its side, with one pole regularly facing the Sun and the other facing away.[47] Its composition is similar to that of Ceres: high in carbon and silicon, and perhaps partially differentiated.[48] Hygiea is a carbonaceous asteroid and, unlike the other largest asteroids, lies relatively close to the plane of the ecliptic.[49]

Rotation

Measurements of the rotation rates of large asteroids in the asteroid belt show that there is an upper limit. No asteroid with a diameter larger than 100 meters has a rotation period smaller than 2.2 hours. For asteroids rotating faster than approximately this rate, the inertia at the surface is greater than the gravitational force, so any loose surface material would be flung out. However, a solid object should be able to rotate much more rapidly. This suggests that most asteroids with a diameter over 100 meters are rubble piles formed through accumulation of debris after collisions between asteroids.[50]

Composition

The physical composition of asteroids is varied and in most cases poorly understood. Ceres appears to be composed of a rocky core covered by an icy mantle, where Vesta is thought to have a nickel-iron core, olivine mantle, and basaltic crust.[51] 10 Hygiea, however, which appears to have a uniformly primitive composition of carbonaceous chondrite, is thought to be the largest undifferentiated asteroid. Most of the smaller asteroids are thought to be piles of rubble held together loosely by gravity, though the largest are probably solid. Some asteroids have moons or are co-orbiting binaries: Rubble piles, moons, binaries, and scattered asteroid families are believed to be the results of collisions that disrupted a parent asteroid.

Asteroids contain traces of amino acids and other organic compounds, and some speculate that asteroid impacts may have seeded the early Earth with the chemicals necessary to initiate life, or may have even brought life itself to Earth. (See also panspermia.)[52] In August 2011, a report, based on NASA studies with meteorites found on Earth, was published suggesting DNA and RNA components (adenine, guanine and related organic molecules) may have been formed on asteroids and comets in outer space.[53] [54] [55]

Only one asteroid, 4 Vesta, which has a reflective surface, is normally visible to the naked eye, and this only in very dark skies when it is favorably positioned. Rarely, small asteroids passing close to Earth may be naked-eye visible for a short time.[56]

Composition is calculated from three primary sources: albedo, surface spectrum, and density. The last can only be determined accurately by observing the orbits of moons the asteroid might have. So far, every asteroid with moons has turned out to be a rubble pile, a loose conglomeration of rock and metal that may be half empty space by volume. The investigated asteroids are as large as 280 km in diameter, and include 121 Hermione (268×186×183 km), and 87 Sylvia (384×262×232 km). Only half a dozen asteroids are larger than 87 Sylvia, though none of them have moons; however, some smaller asteroids are thought to be more massive, suggesting they may not have been disrupted, and indeed 511 Davida, the same size as Sylvia to within measurement error, is estimated to be two and a half times as massive, though this is highly uncertain. The fact that such large asteroids as Sylvia can be rubble piles, presumably due to disruptive impacts, has important consequences for the formation of the Solar system: Computer simulations of collisions involving solid bodies show them destroying each other as often as merging, but colliding rubble piles are more likely to merge. This means that the cores of the planets could have formed relatively quickly.[57]

Surface features

Most asteroids outside the big four (Ceres, Pallas, Vesta, and Hygiea) are likely to be broadly similar in appearance, if irregular in shape. 50-km 253 Mathilde (shown at right) is a rubble pile saturated with craters with diameters the size of the asteroid's radius, and Earth-based observations of 300-km 511 Davida, one of the largest asteroids after the big four, reveal a similarly angular profile, suggesting it is also saturated with radius-size craters.[58] Medium-sized asteroids such as Mathilde and 243 Ida that have been observed up close also reveal a deep regolith covering the surface. Of the big four, Pallas and Hygiea are practically unknown. Vesta has compression fractures encircling a radius-size crater at its south pole but is otherwise a spheroid. Ceres seems quite different in the glimpses Hubble has provided, with surface features that are unlikely to be due to simple craters and impact basins, but details will not be known until *Dawn* arrives in 2015.

253 Mathilde, a C-type asteroid measuring about 50 kilometres (31.07 mi) across, covered in craters half that size. Photograph taken in 1997 by the NEAR Shoemaker probe.

Classification

Asteroids are commonly classified according to two criteria: the characteristics of their orbits, and features of their reflectance spectrum.

Orbital classification

Many asteroids have been placed in groups and families based on their orbital characteristics. Apart from the broadest divisions, it is customary to name a group of asteroids after the first member of that group to be discovered. Groups are relatively loose dynamical associations, whereas families are tighter and result from the catastrophic break-up of a large parent asteroid sometime in the past.[59] Families have only been recognized within the asteroid belt. They were first recognised by Kiyotsugu Hirayama in 1918 and are often called Hirayama families in his honor.

About 30% to 35% of the bodies in the asteroid belt belong to dynamical families each thought to have a common origin in a past collision between asteroids. A family has also been associated with the plutoid dwarf planet Haumea.

Quasi-satellites and horseshoe objects

Some asteroids have unusual horseshoe orbits that are co-orbital with the Earth or some other planet. Examples are 3753 Cruithne and 2002 AA_{29}. The first instance of this type of orbital arrangement was discovered between Saturn's moons Epimetheus and Janus.

Sometimes these horseshoe objects temporarily become quasi-satellites for a few decades or a few hundred years, before returning to their earlier status. Both Earth and Venus are known to have quasi-satellites.

Such objects, if associated with Earth or Venus or even hypothetically Mercury, are a special class of Aten asteroids. However, such objects could be associated with outer planets as well.

Spectral classification

This picture of 433 Eros shows the view looking from one end of the asteroid across the gouge on its underside and toward the opposite end. Features as small as 35 m (115 ft) across can be seen.

In 1975, an asteroid taxonomic system based on colour, albedo, and spectral shape was developed by Clark R. Chapman, David Morrison, and Ben Zellner.[60] These properties are thought to correspond to the composition of the asteroid's surface material. The original classification system had three categories: C-types for dark carbonaceous objects (75% of known asteroids), S-types for stony (silicaceous) objects (17% of known asteroids) and U for those that did not fit into either C or S. This classification has since been expanded to include many other asteroid types. The number of types continues to grow as more asteroids are studied.

The two most widely used taxonomies now used are the Tholen classification and SMASS classification. The former was proposed in 1984 by David J. Tholen, and was based on data collected from an eight-color asteroid survey performed in the 1980s. This resulted in 14 asteroid categories.[61] In 2002, the Small Main-Belt Asteroid Spectroscopic Survey resulted in a modified version of the Tholen taxonomy with 24 different types. Both systems have three broad categories of C, S, and X asteroids, where X consists of mostly metallic asteroids, such as the M-type. There are also several smaller classes.[62]

Note that the proportion of known asteroids falling into the various spectral types does not necessarily reflect the proportion of all asteroids that are of that type; some types are easier to detect than others, biasing the totals.

Problems

Originally, spectral designations were based on inferences of an asteroid's composition.[63] However, the correspondence between spectral class and composition is not always very good, and a variety of classifications are in use. This has led to significant confusion. While asteroids of different spectral classifications are likely to be composed of different materials, there are no assurances that asteroids within the same taxonomic class are composed of similar materials.

At present, the spectral classification based on several coarse resolution spectroscopic surveys in the 1990s is still the standard. Scientists cannot agree on a better taxonomic system, largely due to the difficulty of obtaining detailed measurements consistently for a large sample of asteroids (e.g. finer resolution spectra, or non-spectral data such as densities would be very useful).

Exploration

Until the age of space travel, objects in the asteroid belt were merely pinpricks of light in even the largest telescopes and their shapes and terrain remained a mystery. The best modern ground-based telescopes and the Earth-orbiting Hubble Space Telescope can resolve a small amount of detail on the surfaces of the largest asteroids, but even these mostly remain little more than fuzzy blobs. Limited information about the shapes and compositions of asteroids can be inferred from their light curves (their variation in brightness as they rotate) and their spectral properties, and asteroid sizes can be estimated by timing the lengths of star occulations (when an asteroid passes directly in front of a star). Radar imaging can yield good information about asteroid shapes and orbital and rotational parameters, especially for near-Earth asteroids. In terms of delta-v and propellant requirements, NEOs are more easily accessible than the Moon.[64]

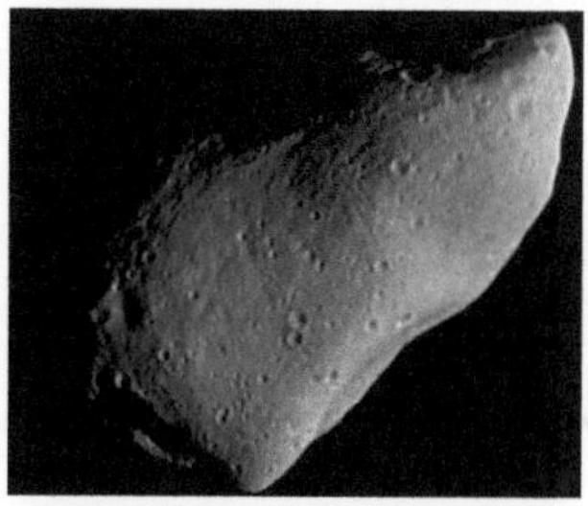

951 Gaspra is the first asteroid to be imaged in close-up.

Vesta, imaged by the Dawn spacecraft

The first close-up photographs of asteroid-like objects were taken in 1971 when the Mariner 9 probe imaged Phobos and Deimos, the two small moons of Mars, which are probably captured asteroids. These images revealed the irregular, potato-like shapes of most asteroids, as did later images from the Voyager probes of the small moons of the gas giants.

The first true asteroid to be photographed in close-up was 951 Gaspra in 1991, followed in 1993 by 243 Ida and its moon Dactyl, all of which were imaged by the Galileo probe en route to Jupiter.

The first dedicated asteroid probe was NEAR Shoemaker, which photographed 253 Mathilde in 1997, before entering into orbit around 433 Eros, finally landing on its surface in 2001.

Other asteroids briefly visited by spacecraft en route to other destinations include 9969 Braille (by Deep Space 1 in 1999), and 5535 Annefrank (by Stardust in 2002).

In September 2005, the Japanese Hayabusa probe started studying 25143 Itokawa in detail and was plagued with difficulties, but returned samples of its surface to earth on June 13, 2010.

The European Rosetta probe (launched in 2004) flew by 2867 Šteins in 2008 and 21 Lutetia, the second-largest asteroid visited to date, in 2010.

In September 2007, NASA launched the Dawn Mission, which orbited the protoplanet 4 Vesta from July 2011 to September 2012, and is planned to orbit 1 Ceres in 2015. 4 Vesta is the largest asteroid visited to date.

On 13 December 2012, China's lunar orbiter Chang'e 2 flew within 2 miles (3 km) of the asteroid 4179 Toutatis on an extended mission.

The Japan Aerospace Exploration Agency (JAXA) plans to launch around 2015 the improved Hayabusa 2 space probe and to return asteroid samples by 2020. Current target for the mission is the C-type asteroid (162173) 1999 JU_3.

In May 2011, NASA announced the OSIRIS-REx sample return mission to asteroid 1999 RQ36, and is expected to launch in 2016.

It has been suggested that asteroids might be used as a source of materials that may be rare or exhausted on earth (asteroid mining), or materials for constructing space habitats (see Colonization of the asteroids). Materials that are heavy and expensive to launch from earth may someday be mined from asteroids and used for space manufacturing and construction.

Fiction

Asteroids and the asteroid belt are a staple of science fiction stories. Asteroids play several potential roles in science fiction: as places human beings might colonize, resources for extracting minerals, hazards encountered by spaceships traveling between two other points, and as a threat to life on Earth by potential impact.

See also

- BOOTES (Burst Observer and Optical Transient Exploring System)
- Asteroid deflection strategies
- Meanings of asteroid names
- Pronunciation of asteroid names
- List of asteroids named after people
- List of asteroids named after places
- Category:Asteroid groups and families
- Category:Binary asteroids
- Category:Asteroids
- List of minor planets
- List of notable asteroids
- Centaur (minor planet)
- Dwarf planet
- Impact event
- Lost asteroid
- Mesoplanet
- Minor planet
- Minor Planet Center
- Marco Polo (spacecraft)
- Near-Earth object
- Orion asteroid mission
- Pioneer 10 space probe
- Rosetta probe

Notes

[1] "Asteroids" (http://ssd.jpl.nasa.gov/?asteroids). NASA – Jet Propulsion Laboratory. . Retrieved 13 September2010.

[2] Asimov, Isaac, and Dole, Stephen H. *Planets for Man* (New York: Random House, 1964), p.43.

[3] "What Are Asteroids And Comets?" (http://neo.jpl.nasa.gov/faq/#ast). *Near Earth Object Program FAQ*. NASA. Archived (http://web.archive.org/web/20100909210213/http://neo.jpl.nasa.gov/faq/) from the original on 9 September 2010. . Retrieved 2010-09-13.

[4] Gould, B. A. (1852). "On the Symbolic Notation of the Asteroids". *Astronomical Journal* **2**: 80. Bibcode 1852AJ......2...80G. doi:10.1086/100212.

[5] Hilton, James L. (2001-09-17). "When Did the Asteroids Become Minor Planets" (http://aa.usno.navy.mil/faq/docs/minorplanets.php). . Retrieved 2006-03-26.

[6] Encke, J. F. (1854). "Beobachtung der Bellona, nebst Nachrichten über die Bilker Sternwarte". *Astronomische Nachrichten* **38** (9): 143. doi:10.1002/asna.18540380907.

[7] Rümker, G. (1855). "Name und Zeichen des von Herrn R. Luther zu Bilk am 19. April entdeckten Planeten". *Astronomische Nachrichten* **40** (24): 373. doi:10.1002/asna.18550402405.

[8] Luther, R. (1856). "Schreiben des Herrn Dr. R. Luther, Directors der Sternwarte zu Bilk, an den Herausgeber". *Astronomische Nachrichten* **42** (7): 107. Bibcode 1855AN.....42..107L. doi:10.1002/asna.18550420705.

[9] "When did the asteroids become minor planets?" (http://www.usno.navy.mil/USNO/astronomical-applications/astronomical-information-center/minor-planets). Naval Meteorology and Oceanography Command. . Retrieved 2011-11-06.

[10] Except for Pluto and, in the astrological community, for a few outer bodies such as 2060 Chiron

[11] Ceres is the largest asteroid and is now classified as a dwarf planet. All other asteroids are now classified as small Solar System bodies along with comets, centaurs, and the smaller trans-Neptunian objects.

[12] http://books.google.ca/books?id=NAMAAAAAMAAJ&pg=PA316&dq=%22planets%22+asteroids

[13] Seares, Frederick H. (1930). "Address of the Retiring President of the Society in Awarding the Bruce Medal to Professor Max Wolf". *Publ. Astr. Soc. Pacific* **42**: 5–22. Bibcode 1930PASP...42....5S. doi:10.1086/123986+(Dead+links).

[14] Chapman, Mary G. (May 17, 1992). "Carolyn Shoemaker, Planetary Astronomer and Most Successful 'Comet Hunter' To Date" (http://astrogeology.usgs.gov/About/People/CarolynShoemaker). USGS. . Retrieved 2008-04-15.

[15] Yeomans, Don. "Near Earth Object Search Programs" (http://neo.jpl.nasa.gov/programs/). NASA. Archived (http://web.archive.org/web/20080424093951/http://neo.jpl.nasa.gov/programs/) from the original on 24 April 2008. . Retrieved 2008-04-15.

[16] "Minor Planet Discover Sites" (http://www.minorplanetcenter.org/iau/lists/MPDiscSites.html). Archived (http://web.archive.org/web/20100830074958/http://www.minorplanetcenter.org/iau/lists/MPDiscSites.html) from the original on 30 August 2010. . Retrieved 2010-08-24.

[17] "Unusual Minor Planets" (http://www.minorplanetcenter.org/iau/lists/Unusual.html). Archived (http://web.archive.org/web/20100827234833/http://www.minorplanetcenter.org/iau/lists/Unusual.html) from the original on 27 August 2010. . Retrieved 2010-08-24.

[18] Beech, M. (September 1995). "On the Definition of the Term Meteoroid". *Quarterly Journal of the Royal Astronomical Society* **36** (3): 281–284. Bibcode 1995QJRAS..36..281B.

[19] The definition of "small Solar System bodies" says that they "include most of the Solar System asteroids, most trans-Neptunian objects, comets, and other small bodies". *The Final IAU Resolution on the definition of "planet" ready for voting* (http://www.iau2006.org/mirror/www.iau.org/iau0602/index.html) (IAU)

[20] "English Dictionary – Browsing Page P-44" (http://www.hyperdictionary.com/dict-e/p-44.html). HyperDictionary.com. . Retrieved 2008-04-15.

[21] Weissman, Paul R., William F. Bottke, Jr., and Harold F. Levinson. "Evolution of Comets into Asteroids." *Southwest Research Institute, Planetary Science Directorate.* 2002. Web Retrieved 3 Aug. 2010 (http://www.boulder.swri.edu/~hal/PDF/asteroids3.pdf)

[22] "Are Kuiper Belt Objects asteroids?" (http://curious.astro.cornell.edu/question.php?number=601), "Ask an astronomer", Cornell University

[23] "Asteroids and Comets" (http://rst.gsfc.nasa.gov/Sect19/Sect19_22.html), NASA website

[24] "Comet Dust Seems More Asteroidy" (http://www.sciam.com/podcast/episode.cfm?id=ADD0878B-D6C3-3B70-7B5BC373545BB82D) *Scientific American*, January 25, 2008

[25] "Comet samples are surprisingly asteroid-like" (http://space.newscientist.com/channel/solar-system/comets-asteroids/dn13224-comet-samples-are-surprisingly-asteroidlike.html), *New Scientist*, 24 January 2008

[26] For instance, a joint NASA-JPL public-outreach website states:

> "We include Trojans (bodies captured in Jupiter's 4th and 5th Lagrange points), Centaurs (bodies in orbit between Jupiter and Neptune), and trans-Neptunian objects (orbiting beyond Neptune) in our definition of "asteroid" as used on this site, even though they may more correctly be called "minor planets" instead of asteroids."

<http://ssd.jpl.nasa.gov/?asteroids>

[27] Questions and Answers on Planets (http://www.iau.org/public_press/news/release/iau0603/questions_answers/), IAU

[28] "Three new planets may join solar system" (http://space.newscientist.com/article.ns?id=dn9761&feedId=online-news_rss20F53), *New Scientist*, 16 August 2006

[29] Bottke, Durda; Nesvorny, Jedicke; Morbidelli, Vokrouhlicky; Levison (2005). "The fossilized size distribution of the main asteroid belt" (http://astro.mff.cuni.cz/davok/papers/fossil05.pdf). *Icarus* **175**: 111. Bibcode 2005Icar..175..111B. doi:10.1016/j.icarus.2004.10.026. .

[30] Kerrod, Robin (2000). *Asteroids, Comets, and Meteors*. Lerner Publications Co.. ISBN 0-585-31763-1.

[31] William B. McKinnon, 2008, "On The Possibility Of Large KBOs Being Injected Into The Outer Asteroid Belt". (http://adsabs.harvard.edu/abs/2008DPS....40.3803M) *American Astronomical Society,* DPS meeting #40, #38.03

[32] Tedesco, Edward; Metcalfe, Leo (April 4, 2002). "New study reveals twice as many asteroids as previously believed" (http://www.spaceref.com/news/viewpr.html?pid=7925) (Press release). European Space Agency. . Retrieved 2008-02-21.

[33] World Book at NASA (http://www.nasa.gov/worldbook/asteroid_worldbook.html)

[34] Neptune also has a few known trojans, and these are thought to actually be much more numerous than the Jovian trojans. However, they are often included in the trans-Neptunian population rather than counted with the asteroids.

[35] Below 10 metres, these rocks are by convention considered to be meteoroids.

[36] Schmidt, B.; Russell, C. T.; Bauer, J. M.; Li, J.; McFadden, L. A.; Mutchler, M.; Parker, J. W.; Rivkin, A. S.; Stern, S. A.; Thomas, P. C. (2007). "Hubble Space Telescope Observations of 2 Pallas". *American Astronomical Society, DPS meeting #39* **39**: 485. Bibcode 2007DPS....39.3519S.

[37] Pitjeva, E. V. (2004). "Estimations of masses of the largest asteroids and the main asteroid belt from ranging to planets, Mars orbiters and landers" (http://adsabs.harvard.edu/abs/2004cosp.meet.2014P). *35th COSPAR Scientific Assembly. Held 18–25 July 2004, in Paris, France*. pp. 2014. .

[38] Davis 2002, "Asteroids III", cited by Željko Ivezić (http://www.astro.washington.edu/users/ivezic/Astr598/lecture4.pdf)

[39] "Recent Asteroid Mass Determinations" (http://home.earthlink.net/~jimbaer1/astmass.txt). Maintained by Jim Baer. Last updated 2010-12-12. Access date 2011-09-02. The values of Juno and Herculina may be off by as much as 16%, and Euphrosyne by a third. The order of the lower eight may change as better data is acquired, but the values do not overlap with any known asteroid outside these twelve.

[40] Pitjeva, E. V. (2005). "High-Precision Ephemerides of Planets—EPM and Determination of Some Astronomical Constants" (http://iau-comm4.jpl.nasa.gov/EPM2004.pdf) (PDF). *Solar System Research* **39** (3): 184. Bibcode 2005SoSyR..39..176P. doi:10.1007/s11208-005-0033-2. .

[41] "Asteroid Density, Porosity, and Structure" (http://www.lpi.usra.edu/books/AsteroidsIII/pdf/3022.pdf). lpi.usra.edu. . Retrieved 2013-01-03.

[42] "The Final IAU Resolution on the Definition of "Planet" Ready for Voting" (http://www.iau.org/public_press/news/detail/iau0602/). IAU. August 24, 2006. . Retrieved 2007-03-02.

[43] Parker, J. W.; Stern, S. A.; Thomas, P. C.; Festou, M. C.; Merline, W. J.; Young, E. F.; Binzel, R. P.; and Lebofsky, L. A. (2002). "Analysis of the First Disk-resolved Images of Ceres from Ultraviolet Observations with the Hubble Space Telescope". *The Astronomical Journal* **123** (1): 549–557. arXiv:astro-ph/0110258. Bibcode 2002AJ....123..549P. doi:10.1086/338093.

[44] "Asteroid 1 Ceres" (http://www.planetary.org/explore/topics/asteroids_and_comets/ceres.html). *The Planetary Society*. Archived (http://web.archive.org/web/20070929092440/http://planetary.org/explore/topics/asteroids_and_comets/ceres.html) from the original

on 29 September 2007. . Retrieved 2007-10-20.

[45] "Key Stages in the Evolution of the Asteroid Vesta" (http://hubblesite.org/newscenter/newsdesk/archive/releases/1995/20/image/c). *Hubble Space Telescope news release*. 1995. Archived (http://web.archive.org/web/20070930203613/http://hubblesite.org/newscenter/newsdesk/archive/releases/1995/20/image/c) from the original on 30 September 2007. . Retrieved 2007-10-20. Russel, C. T.; *et al.* (2007). "Dawn mission and operations" (http://journals.cambridge.org/action/displayAbstract?fromPage=online&aid=414750). *NASA/JPL*. . Retrieved 2007-10-20.

[46] Burbine, T. H. (July 1994). "Where are the olivine asteroids in the main belt?" (http://articles.adsabs.harvard.edu/cgi-bin/nph-iarticle_query?1994Metic..29..453B&data_type=PDF_HIGH&whole_paper=YES&type=PRINTER&filetype=.pdf). *Meteoritics* **29** (4): 453. Bibcode 1994Metic..29..453B. .

[47] Torppa, J.; *et al.* (1996). "Shapes and rotational properties of thirty asteroids from photometric data". *Icarus* **164** (2): 346–383. Bibcode 2003Icar..164..346T. doi:10.1016/S0019-1035(03)00146-5.

[48] Larson, H. P.; Feierberg, M. A.; and Lebofsky, L. A. (1983). "The composition of asteroid 2 Pallas and its relation to primitive meteorites" (http://adsabs.harvard.edu/abs/1983Icar...56..398L). . Retrieved 2007-10-20.

[49] Barucci, M. A.; *et al.* (2002). "10 Hygiea: ISO Infrared Observations" (http://www.lesia.obspm.fr/~crovisier/biblio/preprint/bar02_icarus.pdf) (PDF). Archived (http://web.archive.org/web/20071128200223/http://www.lesia.obspm.fr/~crovisier/biblio/preprint/bar02_icarus.pdf) from the original on 28 November 2007. . Retrieved 2007-10-21. "Ceres the Planet" (http://www.orbitsimulator.com/gravity/articles/ceres.html). *orbitsimulator.com*. Archived (http://web.archive.org/web/20071011154140/http://orbitsimulator.com/gravity/articles/ceres.html) from the original on 11 October 2007. . Retrieved 2007-10-20.

[50] Rossi, Alessandro (2004-05-20). "The mysteries of the asteroid rotation day" (http://spaceguard.iasf-roma.inaf.it/tumblingstone/issues/current/eng/ast-day.htm). The Spaceguard Foundation. . Retrieved 2007-04-09.

[51] HubbleSite - NewsCenter - Asteroid or Mini-Planet? Hubble Maps the Ancient Surface of Vesta (04/19/1995) - Release Images (http://hubblesite.org/newscenter/archive/releases/1995/20/image/)

[52] Life is Sweet: Sugar-Packing Asteroids May Have Seeded Life on Earth (http://www.space.com/scienceastronomy/planetearth/meteor_sugar_011219.html) Archived (http://web.archive.org/20020124092631/http://web.archive.org/web/20020124092631/http://www.space.com/scienceastronomy/planetearth/meteor_sugar_011219.html) January 24, 2002 at the Wayback Machine, Space.com, 19 December 2001

[53] Callahan, M.P.; Smith, K.E.; Cleaves, H.J.; Ruzica, J.; Stern, J.C.; Glavin, D.P.; House, C.H.; Dworkin, J.P. (11 August 2011). "Carbonaceous meteorites contain a wide range of extraterrestrial nucleobases" (http://www.pnas.org/content/early/2011/08/10/1106493108). PNAS. doi:10.1073/pnas.1106493108. . Retrieved 2011-08-15.

[54] Steigerwald, John (8 August 2011). "NASA Researchers: DNA Building Blocks Can Be Made in Space" (http://www.nasa.gov/topics/solarsystem/features/dna-meteorites.html). NASA. . Retrieved 2011-08-10.

[55] ScienceDaily Staff (9 August 2011). "DNA Building Blocks Can Be Made in Space, NASA Evidence Suggests" (http://www.sciencedaily.com/releases/2011/08/110808220659.htm). ScienceDaily. . Retrieved 2011-08-09.

[56] Closest Flyby of Large Asteroid to be Naked-Eye Visible (http://www.space.com/spacewatch/050204_2004_mn4.html), Space.com, 4 February 2005

[57] Marchis, Descamps, et al. *Icarus*, Feb. 2011

[58] A.R. Conrad *et al.* 2007. " Direct measurement of the size, shape, and pole of 511 Davida with Keck AO in a single night (http://www2.keck.hawaii.edu/inst/people/conrad/research/pub/d511.pdf)", *Icarus,* doi:10.1016/j.icarus.2007.05.004

[59] Zappalà, V. (1995). "Asteroid families: Search of a 12,487-asteroid sample using two different clustering techniques". *Icarus* **116** (2): 291–314. Bibcode 1995Icar..116..291Z. doi:10.1006/icar.1995.1127.

[60] Chapman, C. R. (1975). "Surface properties of asteroids: A synthesis of polarimetry, radiometry, and spectrophotometry". *Icarus* **25** (1): 104–130. Bibcode 1975Icar...25..104C. doi:10.1016/0019-1035(75)90191-8.

[61] Tholen, D. J. (March 8–11, 1988). "Asteroid taxonomic classifications" (http://adsabs.harvard.edu/abs/1989aste.conf.1139T). *Asteroids II; Proceedings of the Conference*. Tucson, AZ: University of Arizona Press. pp. 1139–1150. . Retrieved 2008-04-14.

[62] Bus, S. J. (2002). "Phase II of the Small Main-belt Asteroid Spectroscopy Survey: A feature-based taxonomy". *Icarus* **158** (1): 146. Bibcode 2002Icar..158..146B. doi:10.1006/icar.2002.6856.

[63] McSween Jr., Harry Y. (1999). *Meteorites and their Parent Planets* (2nd ed.). Oxford University Press. ISBN 0-521-58751-4.

[64] A Piloted Orion Flight to a Near-Earth Object: A Feasibility Study (http://ti.arc.nasa.gov/m/project/neo/pdf/NEO_feasibility.pdf)

References

External links

- Rocks from the Main Belt asteroids (http://rocksfromspace.open.ac.uk/asteroid_belt_detail.htm)
- Alphabetical list of minor planet names (ASCII) (http://www.minorplanetcenter.org/iau/lists/MPNames.html) (Minor Planet Center)
- Near Earth Asteroid Tracking (NEAT) (http://neat.jpl.nasa.gov/)
- Asteroids Page (http://solarsystem.nasa.gov/planets/profile.cfm?Object=Asteroids) at NASA's Solar System Exploration (http://solarsystem.nasa.gov/)
- Asteroid Simulator with Moon and Earth (http://www.colorado.edu/physics/2000/applets/satellites.html)
- Alphabetical and numerical lists of minor planet names (Unicode) (http://www.ipa.nw.ru/PAGE/DEPFUND/LSBSS/englenam.htm) (Institute of Applied Astronomy)
- Future Asteroid Interception Research (http://www.fair-society.org/)
- Near Earth Objects Dynamic Site (http://newton.dm.unipi.it/cgi-bin/neodys/neoibo)
- Asteroids Dynamic Site (http://hamilton.dm.unipi.it/astdys/) Up-to-date osculating orbital elements and proper orbital elements University of Pisa, Italy.
- JPL small bodies database (http://ssd.jpl.nasa.gov/?sb_elem) Current down-loadable ASCII table of orbit data and absolute mags H for over 200000 asteroids, sorted by number. Caltech/JPL.
- Asteroid naming statistics (http://quasar.ipa.nw.ru/PAGE/DEPFUND/LSBSS/statmpn.htm)
- Spaceguard UK (http://www.spaceguarduk.com/)
- Committee on Small Body Nomenclature (http://www.ss.astro.umd.edu/IAU/csbn/)
- List of minor planet orbital groupings and families from ProjectPluto (http://www.projectpluto.com/mp_group.htm)
- Cunningham, Clifford, "Introduction to Asteroids: The Next Frontier", ISBN 0-943396-16-6
- James L. Hilton: When Did the Asteroids Become Minor Planets? (http://aa.usno.navy.mil/faq/docs/minorplanets.php)
- Kirkwood, Daniel; *Relations between the Motions of some of the Minor Planets* (1874). (http://adsabs.harvard.edu/cgi-bin/nph-bib_query?bibcode=1874MNRAS..35...61K&db_key=AST&high=40daf3f6f901929)
- Schmadel, L.D. (2003). *Dictionary of Minor Planet Names.* 5th ed. IAU/Springer-Verlag: Heidelberg.
- Asteroid articles in Planetary Science Research Discoveries (http://www.psrd.hawaii.edu/Archive/Archive-Asteroids.html)
- Catalogue of the Solar System Small Bodies Orbital Evolution (http://smallbodies.ru/en/)
- TECA Table of next close approaches to the Earth (http://www.brera.mi.astro.it/sormano/teca.html)
- SAEL Small Asteroids Encounter List (http://www.brera.mi.astro.it/sormano/sael.html)
- MBPL Minor Body Priority List (http://www.brera.mi.astro.it/sormano/mbpl.html)
- PCEL Planetary Close Encounter List (http://www.brera.mi.astro.it/sormano/pcel.html)

Trojan (astronomy)

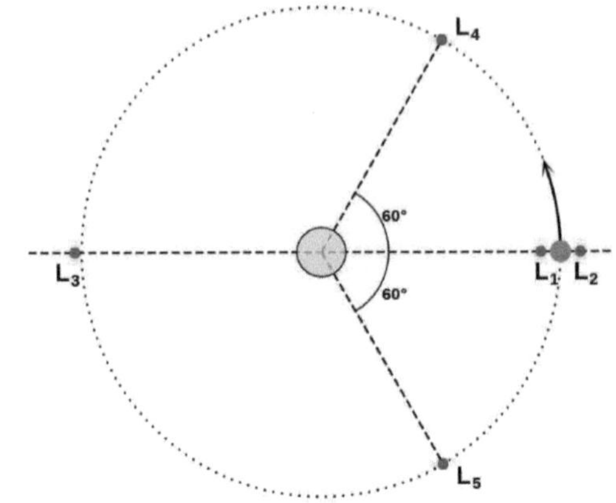

The **trojan points** are those labelled L_4 and L_5, highlighted in red, on the orbital path of the secondary object (blue), around the primary object (yellow).

In astronomy, a **trojan** is a minor planet or natural satellite (moon) that shares an orbit with a planet or larger moon, but does not collide with it because it orbits around one of the two Lagrangian points of stability (trojan points), L_4 and L_5, which lie approximately 60° ahead of and behind the larger body, respectively. It is also sometimes called a **Lagrangian** object. Trojan objects are one type of co-orbital object. In this arrangement, the massive star and the smaller planet orbit about their common barycenter—a location in space where the forces of their mutual gravitational attraction balance each other out. A much smaller mass located at one of the Lagrangian points is subject to a combined gravitational force that acts through this barycenter. Hence the object can orbit around the barycenter with the same orbital period as the planet, and the arrangement can remain stable over time.[1]

Trojan asteroids are asteroids that orbit near one of the two trojan points of a planet. A trojan moon is a moon residing at the trojan point of another (larger) moon. Trojan planets are theoretical planets that orbit at the trojan points of other planets.

Saturn has the only known trojan satellites: Saturn's moon Tethys has two trojan moons (Telesto and Calypso), and Dione also has two trojan moons (Helene and Polydeuces).

In 2011, NASA announced the discovery of the first known Earth trojan.[2]

Numerical calculations of the orbital dynamics involved indicate that Saturn and Uranus probably do not have any primordial trojan asteroids.[3]

Trojan asteroids

In 1772, the Italian–French mathematician and astronomer Joseph-Louis Lagrange obtained two constant-pattern solutions (collinear and equilateral) of the general three-body problem.[4] In the restricted three-body problem, with one mass negligible (which Lagrange did not consider), the five possible positions of that mass are now termed Lagrangian points.

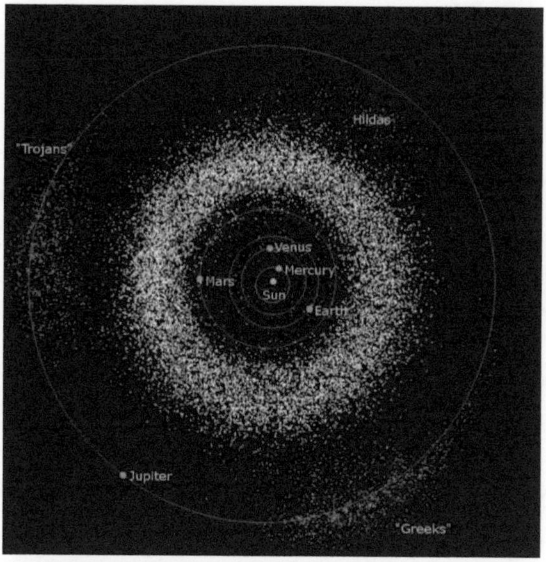

Trojan asteroids of Jupiter (coloured green) in front of and behind the planet along its orbital path. Also shown is the asteroid belt between the orbits of Mars and Jupiter (white), and the Hilda family of asteroids (brown).

The term "trojan" originally referred to the "trojan asteroids" that orbit close to the Lagrangian points of Jupiter. These have long been named after characters from the Trojan War of Greek mythology. By convention, the asteroids orbiting near the L_4 point of Jupiter are named after the characters from the Greek side of the war, whereas those orbiting near the L_5 of Jupiter are from the Trojan side. There are two exceptions that were named before the convention was put in place, the Greek-themed 617 Patroclus and the Trojan-themed 624 Hektor, which were assigned to the wrong sides.[5]

Astronomers estimate that the Jupiter trojans are about as numerous as the asteroids of the asteroid belt.[6]

Later on, objects were found orbiting near the Lagrangian points of Neptune, Mars, and Earth.[7] Asteroids at the Lagrangian points of planets other than Jupiter may be called Lagrangian asteroids.[8]

- Three Mars trojans are known: 5261 Eureka, 1998 VF_{31}, and 1999 UJ_7.[9] 2007 NS_2 is possibly one.
- Nine Neptune trojans[10] are known, but they have been estimated to outnumber the Jupiter trojans by an order of magnitude. [11] [12]
- 2010 TK_7 was confirmed to be the first known Earth trojan in 2011. It is located in the L_4 Lagrangian point, which lies ahead of Earth.[13]

See also

- Lissajous orbit
- List of objects at Lagrangian points

References

[1] Robutel, Philippe; Souchay, Jean (2010), "An introduction to the dynamics of trojan asteroids" (http://books.google.com/books?id=CLUYgQlWz4IC&pg=PA197), in Dvorak, Rudolf; Souchay, Jean, *Dynamics of Small Solar System Bodies and Exoplanets*, Lecture Notes in Physics, **790**, Springer, p. 197, ISBN 3-642-04457-3,

[2] NASA's WISE Mission Finds First Trojan Asteroid Sharing Earth's Orbit 7.27.11 (http://www.nasa.gov/mission_pages/WISE/news/wise20110727.html)

[3] Sheppard, Scott S.; Trujillo, Chadwick A. (June 2006). "A Thick Cloud of Neptune Trojans and their Colors" (http://www.dtm.ciw.edu/users/sheppard/pub/Sheppard06NepTroj.pdf) (PDF). *Science* **313** (5786): 511–514. Bibcode 2006Sci...313..511S. doi:10.1126/science.1127173. PMID 16778021. . Retrieved 2008-02-26.

[4] *Essai sur le Problème des Trois Corps* (http://www.ltas-vis.ulg.ac.be/cmsms/uploads/File/Lagrange_essai_3corps.pdf)

[5] Wright, Alison (August 1, 2011). "Planetary science: The Trojan is out there" (http://www.nature.com/nphys/journal/v7/n8/full/nphys2061.html). *Nature Physics* **7** (8): 592. Bibcode 2011NatPh...7..592W. doi:10.1038/nphys2061. . Retrieved 2011-08-12.

[6] Yoshida, Fumi; Nakamura, Tsuko (2005). "Size distribution of faint L_4 Trojan asteroids". *The Astronomical journal* **130** (6): 2900–11. Bibcode 2005AJ....130.2900Y. doi:10.1086/497571.

[7] Connors, Martin; Wieger, Paul; Veillet, Christian (27 July 2011). "Earth's Trojan asteroid" (http://www.nature.com/nature/journal/v475/n7357/full/nature10233.html). *Nature* **475** (7357): 481–483. Bibcode 2011Natur.475..481C. doi:10.1038/nature10233. PMID 21796207. . Retrieved 2011-07-27.

[8] Whiteley, Robert J.; and Tholen, David J.; "A CCD Search for Lagrangian Asteroids of the Earth–Sun System", *Icarus* 136:1, November 1998:154–167

[9] "List of Martian Trojans" (http://www.minorplanetcenter.org/iau/lists/MarsTrojans.html). . Retrieved 2012-01-31.

[10] "List of Neptune Trojans" (http://www.minorplanetcenter.org/iau/lists/NeptuneTrojans.html). . Retrieved 2012-12-18.

[11] Chiang, Eugene I. & Lithwick, Yoram; *Neptune Trojans as a Testbed for Planet Formation*, The Astrophysical Journal, **628**, pp. 520–532 Preprint (http://www.arxiv.org/abs/astro-ph/0502276)

[12] Powell, David (30 January 2007). "Neptune May Have Thousands of Escorts" (http://www.space.com/scienceastronomy/070130_st_neptune_trojans.html). Space.com. . Retrieved 2007-03-08.

[13] Choi, Charles Q. (27 July 2011). "First Asteroid Companion of Earth Discovered at Last" (http://www.space.com/12443-earth-asteroid-companion-discovered-2010-tk7.html). Space.com. . Retrieved 2011-07-27.

Centaur (minor planet)

Centaurs are small Solar System bodies with a semi-major axis between those of the giant planets. They therefore have unstable orbits that cross or have crossed the orbits of one or more of the giant planets, and have dynamic lifetimes of a few million years.[1] Centaurs typically behave with characteristics of both asteroids and comets. They are named after the mythological race of beings, centaurs, which were a mixture of horse and human. It has been estimated that there are around 44,000 centaurs in the Solar System with diameters larger than 1 km.[1]

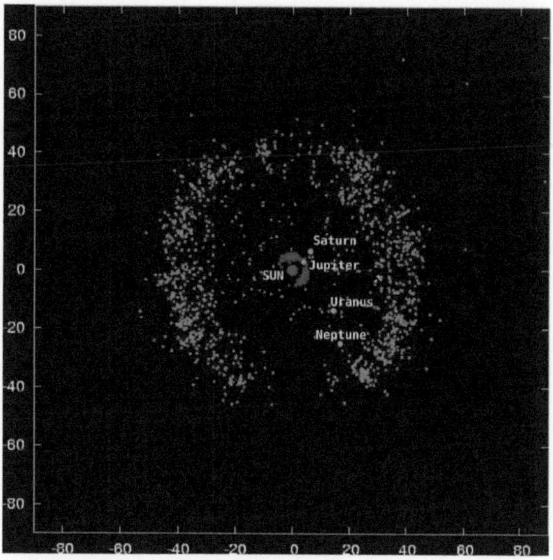

Positions of known outer Solar System objects. The centaurs are those objects (in orange) that lie generally **inwards** of the Kuiper belt (in green) and outside the Jupiter Trojans (pink).

The first centaur to be discovered was 944 Hidalgo in 1920. However, they were not recognized as a distinct population until the discovery of 2060 Chiron in 1977. The largest known centaur is 10199 Chariklo, discovered in 1997, which at 260 km in diameter is as big as a mid-sized main-belt asteroid.

No centaur has been photographed up close, although there is evidence that Saturn's moon Phoebe, imaged by the Cassini probe in 2004, may be a captured centaur. In addition, the Hubble Space Telescope has gleaned some information about the surface features of 8405 Asbolus.

As of 2008, three centaurs have been found to display cometary comas: Chiron, 60558 Echeclus, and 166P/NEAT. Chiron and Echeclus are therefore classified as both asteroids and comets. Other centaurs such as 52872 Okyrhoe are suspected of showing cometary activity. Any centaur that is perturbed close enough to the Sun is expected to become a comet.

Classification

The generic definition of a centaur is a small body that orbits the Sun between Jupiter and Neptune and crosses the orbits of one or more of the giant planets. Due to the inherent long-term instability of orbits in this region, even centaurs such as 2000 GM_{137} and 2001 XZ_{255}, which do not currently cross the orbit of any planet, are in gradually changing orbits that will be perturbed until they start to cross the orbit of one or more of the giant planets.[1]

However, different institutions have different criteria for classifying borderline objects, based on particular values of their orbital elements:

- The Minor Planet Center (MPC) defines centaurs as having a perihelion beyond the orbit of Jupiter and a semi-major axis less than that of Neptune.[2]

- The Jet Propulsion Laboratory (JPL) similarly defines centaurs as having a semi-major axis, *a*, between those of Jupiter and Neptune (5.5 AU < a < 30.1 AU).[3]
- In contrast, the Deep Ecliptic Survey (DES) defines centaurs using a dynamical classification scheme. These classifications are based on the simulated change in behavior of the present orbit when extended over 10 million years. The DES defines centaurs as non-resonant objects whose instantaneous (osculating) perihelia are less than the osculating semi-major axis of Neptune at any time during the simulation. This definition is intended to be synonymous with planet-crossing orbits and to suggest comparatively short lifetimes in the current orbit.[4]

The collection *The Solar System Beyond Neptune* (2008) uses the traditional definition of centaurs, limited to semi-major axes smaller than that of Neptune, classifying the objects on unstable orbits beyond this limit as members of the scattered disk.[5] Yet, other astronomers still prefer to define centaurs as objects that are non-resonant with a perihelion inside the orbit of Neptune that can be shown to likely cross the Hill sphere of a gas giant within the next 10 million years.[6] Thus centaurs can be thought of as inward scattered objects that interact more aggressively and scatter more quickly than typical scattered disc objects.

The JPL Small-Body Database lists 183 centaurs.[7] There are an additional 40 trans-Neptunian objects with a semi-major axis further than Neptune (a > 30.1 AU) and perihelion closer than the orbit of Uranus (q < 19 AU).[7] The Committee on Small Body Nomenclature of the International Astronomical Union has not formally weighed in on either side of the debate. Instead, it has adopted the following naming convention for such objects: befitting their centaur-like transitional orbits between TNOs and comets, "objects on unstable, non-resonant, giant-planet-crossing orbits with semimajor axes greater than Neptune's" are to be named for other hybrid and shape-shifting mythical creatures. Thus far, only the binary objects Ceto and Phorcys and Typhon and Echidna have been named according to the new policy.[8]

Other objects caught between these differences in classification methods include (44594) 1999 OX_3, which has a semi-major axis of 32 AU but crosses the orbits of both Uranus and Neptune. Among the inner centaurs, 2005 VD, with a perihelion distance very near Jupiter, is listed as a centaur by both JPL and DES.

Centaurs with measured diameters listed as possible dwarf planets according to Mike Brown's website include 10199 Chariklo, 2060 Chiron, and 54598 Bienor.[9]

Orbits

Distribution

The diagram at right illustrates the orbits of all known centaurs in relation to the orbits of the planets. For selected objects, the eccentricity of the orbits is represented by red segments (extending from perihelion to aphelion).

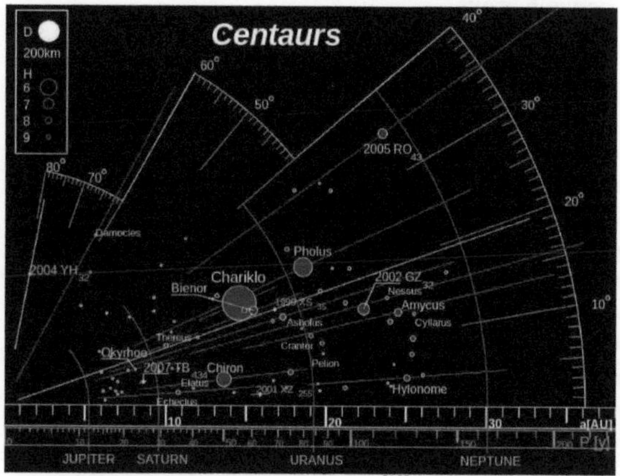

Orbits of known centaurs[10]

Centaurs' orbits are characterised by a wide range of eccentricity, from highly eccentric (Pholus, Asbolus, Amicus, Nessus) to more circular (Chariklo and the *Saturn-crossers*: Thereus, Okyrhoe).

To illustrate the range of the orbits' parameters, a few objects with very unusual orbits are plotted in yellow on the diagram:

- 1999 XS_{35} (Apollo asteroid) follows an extremely eccentric orbit (e=0.947), leading it from inside Earth's orbit (0.94 AU) to well beyond Neptune (>34 AU)
- 2007 TB_{434} follows a quasi-circular orbit (e<0.026)
- 2001 XZ_{255} has the lowest inclination (i<3°).
- Damocles is among a few centaurs on orbits with extreme inclination (prograde i>70°, e.g. 2007 DA_{61}, 2004 YH_{32}, retrograde i<120° e.g. 2005 JT_{50}; not shown)
- 2004 YH_{32} follows such a highly inclined orbit (nearly 80°) that, while it crosses from the distance of the asteroid belt from the Sun to past the distance of Saturn, it does not even cross Jupiter relative to the plane of Jupiter's orbit.

A dozen known centaurs, including Dioretsa ("asteroid" spelled backwards), follow retrograde orbits.

Changing orbits

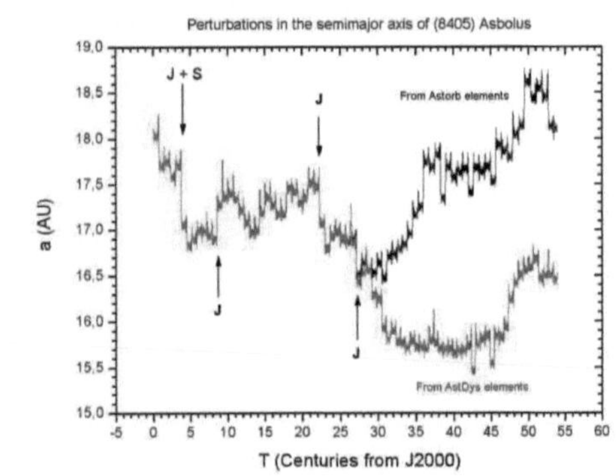

The semi-major axis of Asbolus during the next 5500 years. After the Jupiter encounter of year 4713 (27 Cy in the future) the two orbits become substantially divergent.[11]

Because the centaurs cross the orbits of the giant planets and are not protected by orbital resonances, their orbits are unstable within a timescale of 10^6–10^7 years.[12] For example, 55576 Amycus is in an unstable orbit near the 3:4 resonance of Uranus.[1] Dynamical studies of their orbits indicate that centaurs are probably an intermediate orbital state of objects transitioning from the Kuiper belt to the Jupiter family of short-period comets. Objects may be perturbed from the Kuiper belt, whereupon they become Neptune-crossing and interact gravitationally with that planet (see theories of origin). They then become classed as centaurs, but their orbits are chaotic, evolving relatively rapidly as the centaur makes repeated close approaches to one or more of the outer planets. Some centaurs will evolve into Jupiter-crossing orbits whereupon their perihelia may become reduced into the inner Solar System and they may be reclassified as active comets in the Jupiter family if they display cometary activity. Centaurs will thus ultimately collide with the Sun or a planet or else they may be ejected into interstellar space after a close approach to one of the planets, particularly Jupiter.

Physical characteristics

The relatively small size of centaurs precludes surface observations, but colour indices and spectra can indicate possible surface composition and can provide insight into the origin of the bodies.[12]

Colours

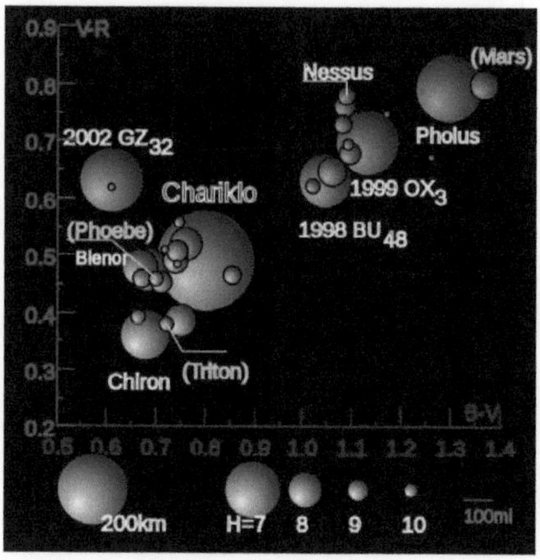

Colour distribution of centaurs

Centaurs display a puzzling diversity of colour that challenges any simple model of surface composition.[13] In the side-diagram, the colour indices are measures of apparent magnitude of an object through blue (B), visible (V) *i.e.* green-yellow and red (R) filters. The diagram illustrates these differences (in enhanced colour) for all centaurs with known colour indices. For reference, two moons: Triton and Phoebe, and planet Mars are plotted (yellow labels, size not to scale).

Centaurs appear to be grouped into two classes:

- very red, for example 5145 Pholus
- blue (or blue-grey, according to some authors), for example 2060 Chiron

There are numerous theories to explain this colour difference, but they can be divided broadly into two categories:

- The colour difference results from a difference in the origin and/or composition of the centaur (see origin below)
- The colour difference reflects a different level of space-weathering from radiation and/or cometary activity.

As examples of the second category, the reddish colour of Pholus has been explained as a possible mantle of irradiated red organics, whereas Chiron has instead had its ice exposed due to its periodic cometary activity, giving it a blue/grey index. The correlation with activity and color is not certain, however, as the active centaurs span the range of colors from blue (Chiron) to red (166P/NEAT).[14] Alternatively, Pholus may have been only recently expelled from the Kuiper belt, so that surface transformation processes have not yet taken place.

A. Delsanti *et al.* suggest multiple competing processes: reddening by the radiation, and blushing by collisions.[15][16]

Spectra

The interpretation of spectra is often ambiguous, related to particle sizes and other factors, but the spectra offer an insight into surface composition. As with the colours, the observed spectra can fit a number of models of the surface.

Water ice signatures have been confirmed on a number of centaurs[12] (including 2060 Chiron, 10199 Chariklo and 5145 Pholus). In addition to the water ice signature, a number of other models have been put forward:

- Chariklo's surface has been suggested to be a mixture of tholins (like those detected on Titan and Triton) with amorphous carbon.
- Pholus has been suggested to be covered by a mixture of Titan-like tholins, carbon black, olivine[17] and methanol ice.
- The surface of 52872 Okyrhoe has been suggested to be a mixture of kerogens, olivines and small percentage of water ice.
- 8405 Asbolus has been suggested to be a mixture of 15% Triton-like tholins, 8% Titan-like tholin, 37% amorphous carbon and 40% ice tholin.

Chiron, the only centaur with known cometary activity, appears to be the most complex. The spectra observed vary depending on the period of the observation. Water ice signature was detected during a period of low activity and disappeared during high activity.[19] [20] [21]

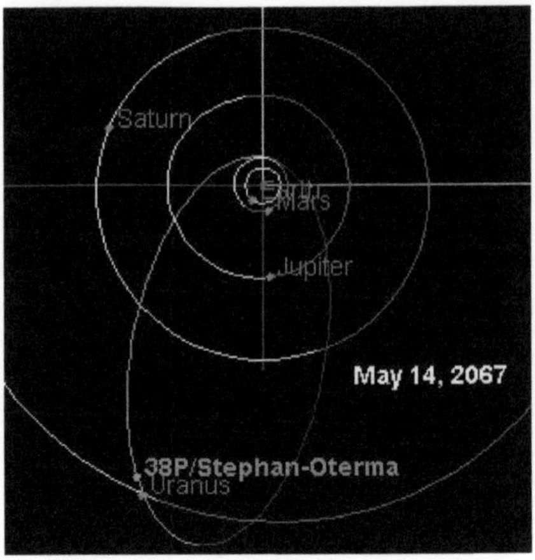

Comet 38P exhibits centaur-like behavior by making close approaches to Jupiter, Saturn, and Uranus between 1982 and 2067.[18]

Similarities to comets

Observations of Chiron in 1988 and 1989 near its perihelion found it to display a coma (a cloud of gas and dust evaporating from its surface). It is thus now officially classified as both a comet and an asteroid, although it is far larger than a typical comet and there is some lingering controversy. Other centaurs are being monitored for comet-like activity: so far two, 60558 Echeclus, and 166P/NEAT have shown such behavior. 166P/NEAT was discovered while it exhibited a coma, and so is classified as a comet, though its orbit is that of a centaur. 60558 Echeclus was discovered without a coma but recently became active,[22] and so it is now accordingly also classified as both a comet and an asteroid.

There is no clear orbital distinction between centaurs and comets. Both 29P/Schwassmann-Wachmann and 39P/Oterma have been referred to as centaurs since they have typical centaur orbits. The comet 39P/Oterma is currently inactive and was seen to be active only before it was perturbed into a centaur orbit by Jupiter in 1963.[23] The faint comet 38P/Stephan–Oterma would probably not show a coma if it had a perihelion distance beyond Jupiter's orbit at 5 AU. By the year 2200, comet 78P/Gehrels will probably migrate outwards into a centaur-like orbit.

Theories of origin

The study of centaur development is rich in recent developments but still hampered by limited physical data. Different models have been put forward for possible origin of centaurs.

Simulations indicate that the orbit of some Kuiper-belt objects can be perturbed, resulting in the object's expulsion so that it becomes a centaur. Scattered disk objects would be dynamically the best candidates[24] for such expulsions, but their colours do not fit the bicoloured nature of the centaurs. Plutinos are a class of Kuiper-belt object that display a similar bicoloured nature, and there are suggestions that not all plutinos' orbits are as stable as initially thought, due to perturbation by Pluto.[25] Further developments are expected with more physical data on KBOs.

Notable centaurs

Well-known centaurs include:

Name	Year	Discoverer	Half-life[1] (forward)	Class
55576 Amycus	2002	NEAT at Palomar	11.1 Myr	UE
54598 Bienor	2000			
10370 Hylonome	1995	Mauna Kea Observatory	6.3 Myr	UN
10199 Chariklo	1997	Spacewatch	10.3 Myr	U
8405 Asbolus	1995	Spacewatch (James V. Scotti)	0.86 Myr	SN
7066 Nessus	1993	Spacewatch (David L. Rabinowitz)	4.9 Myr	SE
5145 Pholus	1992	Spacewatch (David L. Rabinowitz)	1.28 Myr	SN
2060 Chiron	1977	Charles T. Kowal	1.03 Myr	SU

Notes

[1] Horner, J.; Evans, N.W.; Bailey, M. E. (2004). "Simulations of the Population of Centaurs I: The Bulk Statistics". *Monthly Notices of the Royal Astronomical Society* **354** (3): 798–810. arXiv:astro-ph/0407400. Bibcode 2004MNRAS.354..798H. doi:10.1111/j.1365-2966.2004.08240.x.

[2] "Unusual Minor Planets" (http://www.minorplanetcenter.org/iau/lists/Unusual.html). Minor Planet Center. . Retrieved 2010-10-25.

[3] "Orbit Classification (Centaur)" (http://ssd.jpl.nasa.gov/sbdb_help.cgi?class=CEN). JPL Solar System Dynamics. . Retrieved 2008-10-13.

[4] Elliot, J.L.; Kern, Buie, Trilling; et al. (2005). "The Deep Ecliptic Survey: A Search for Kuiper Belt Objects and Centaurs. II. Dynamical Classification, the Kuiper Belt Plane, and the Core Population" (http://www.iop.org/EJ/abstract/1538-3881/129/2/1117). *The Astronomical Journal* **129** (2): 1117–1162. Bibcode 2005AJ....129.1117E. doi:10.1086/427395. . Retrieved 2008-09-22.

[5] B. Gladman, B. Marsden, C. VanLaerhoven (2008). "Nomenclature in the Outer Solar System". *In **The Solar System Beyond Neptune**, ISBN 978-0-8165-2755-7.*

[6] Chaing, Eugene; Buie, Grundy, Holman; et al. (2007). "A Brief History of Transneptunian Space". *Protostars and Planets V, B. Reipurth, D. Jewitt, and K. Keil (eds.), University of Arizona Press, Tucson*: 895–911. arXiv:astro-ph/0601654. Bibcode 2006astro.ph..1654C.

[7] "JPL Small-Body Database Search Engine" (http://ssd.jpl.nasa.gov/sbdb_query.cgi). JPL Solar System Dynamics. . Retrieved 2010-12-27.

[8] Grundy, Will; Stansberry, J.A.; Noll, K; Stephens, D.C.; Trilling, D.E.; Kern, S.D.; Spencer, J.R.; Cruikshank, D.P. et al. (2007). "The orbit, mass, size, albedo, and density of (65489) Ceto/Phorcys: A tidally-evolved binary Centaur". *Icarus* **191** (1): 286–297. arXiv:0704.1523. Bibcode 2007Icar..191..286G. doi:10.1016/j.icarus.2007.04.004.

[9] Michael E. Brown. "How many dwarf planets are there in the outer solar system? (updates daily)" (http://www.gps.caltech.edu/~mbrown/dps.html). California Institute of Technology. . Retrieved 2012-01-16.

[10] For the purpose of this diagram, an object is classified as a centaur if its semi-major axis lies between Jupiter and Neptune. Last update: October 2008

[11] "Three clones of Centaur 8405 Asbolus making passes within 450Gm" (http://home.surewest.net/kheider/astro/AsbolusClones.txt). . Retrieved 2009-05-02. (Solex 10) (http://chemistry.unina.it/~alvitagl/solex/)

[12] Jewitt, David C.; A. Delsanti (2006). "The Solar System Beyond The Planets". *Solar System Update : Topical and Timely Reviews in Solar System Sciences*. Springer-Praxis Ed.. ISBN 3-540-26056-0. (Preprint version (pdf) (http://www.ifa.hawaii.edu/faculty/jewitt/papers/2006/DJ06.pdf))

[13] M. A. Barucci, A. Doressoundiram, and D. P. Cruikshank, "Physical Characteristics of TNOs and Centaurs" (2003), available on the web (http://www.lesia.obspm.fr/~alaind/TNO/Barucci2003_comet2.pdf) (accessed 3/20/2008)

[14] Bauer, J. M., Fernández, Y. R., & Meech, K. J. 2003. " An Optical Survey of the Active Centaur C/NEAT (2001 T4) (http://www.journals.uchicago.edu/PASP/journal/issues/v115n810/203101/203101.html)", Publication of the Astronomical Society of the Pacific", **115**, 981

[15] Peixinho, N.; Doressoundiram, A.; Delsanti, A.; Boehnhardt, H.; Barucci, M. A.; Belskaya, I. (2003). "Reopening the TNOs Color Controversy: Centaurs Bimodality and TNOs Unimodality". *Astronomy and Astrophysics* **410** (3): L29–L32. arXiv:astro-ph/0309428. Bibcode 2003A&A...410L..29P. doi:10.1051/0004-6361:20031420.

[16] Hainaut & Delsanti (2002) *Color of Minor Bodies in the Outer Solar System* Astronomy & Astrophysics, **389**, 641 datasource (http://www.sc.eso.org/~ohainaut/MBOSS)

[17] A class of Magnesium Iron Silicates (Mg, Fe)$_2$SiO$_4$, common components of igneous rocks.

[18] "JPL Close-Approach Data: 38P/Stephan-Oterma" (http://ssd.jpl.nasa.gov/sbdb.cgi?sstr=38P;cad=1#cad). 1981-04-04 last obs. . Retrieved 2009-05-07.

[19] Dotto, E; Barucci, M A; De Bergh, C, *Colours and composition of the centaurs*, Earth, Moon, and Planets, **92**, no. 1–4, pp. 157–167. (June 2003)

[20] Luu, Jane X.; Jewitt, David; Trujillo, C. A. (2000). "Water Ice on 2060 Chiron and its Implications for Centaurs and Kuiper Belt Objects". *The Astrophysical Journal* **531** (2): L151–L154. arXiv:astro-ph/0002094. Bibcode 2000ApJ...531L.151L. doi:10.1086/312536. PMID 10688775.

[21] Fernandez, Y. R.; Jewitt, D. C.; Sheppard, S. S. (2002). "Thermal Properties of Centaurs Asbolus and Chiron". *The Astronomical Journal* **123** (2): 1050–1055. arXiv:astro-ph/0111395. Bibcode 2002AJ....123.1050F. doi:10.1086/338436.

[22] Y-J. Choi, P.R. Weissman, and D. Polishook *(60558) 2000 EC_98*, IAU Circ., **8656** (Jan. 2006), 2.

[23] Mazzotta Epifani, E.; Palumbo; Capria; Cremonese;; et al. (2006). "The dust coma of the active Centaur P/2004 A1 (LONEOS): a CO-driven environment?" (http://google.com/search?q=cache:www.aanda.org/articles/aa/ps/2006/48/aa5189-06.ps.gz). *Astronomy & Astrophysics* **460** (3): 935–944. Bibcode 2006A&A...460..935M. doi:10.1051/0004-6361:20065189. . Retrieved 2009-05-08.

[24] for instance, the centaurs could be part of an "inner" scattered disc of objects perturbed inwards from the Kuiper belt (http://www.zanestein.com/page4_1.htm).

[25] Wan, X.-S; Huang, T.-Y. (2001). "The orbit evolution of 32 plutinos over 100 million year". *Astronomy and Astrophysics* **368** (2): 700–705. Bibcode 2001A&A...368..700W. doi:10.1051/0004-6361:20010056.

See also

- Asteroid
- Dwarf planet
- Minor planet

References

External links

- List of Centaurs and Scattered-Disk Objects (http://www.minorplanetcenter.org/iau/lists/Centaurs.html)
- Centaurs from The Encyclopedia of Astrobiology Astronomy and Spaceflight (http://www.daviddarling.info/encyclopedia/C/Centaur.html)
- Horner, Jonathan; Lykawka, Patryk Sofia (2010). "Planetary Trojans – the main source of short period comets?". *International Journal of Astrobiology* **9** (04): 227–234. arXiv:1007.2541. Bibcode 2010IJAsB...9..227H. doi:10.1017/S1473550410000212.

Kuiper belt

The **Kuiper belt** /ˈkaɪpər/ (rhymes with "viper"), sometimes called the **Edgeworth–Kuiper belt**, is a region of the Solar System beyond the planets, extending from the orbit of Neptune (at 30 AU) to approximately 50 AU from the Sun.[1] It is similar to the asteroid belt, but it is far larger—20 times as wide and 20 to 200 times as massive.[2] [3] Like the asteroid belt, it consists mainly of small bodies, or remnants from the Solar System's formation. While most asteroids are composed primarily of rock and metal, most Kuiper belt objects are composed largely of frozen volatiles (termed "ices"), such as methane, ammonia and water. The classical belt is home to at least three dwarf planets: Pluto, Haumea, and Makemake. Some of the Solar System's moons, such as Neptune's Triton and Saturn's Phoebe, are also believed to have originated in the region.[4] [5]

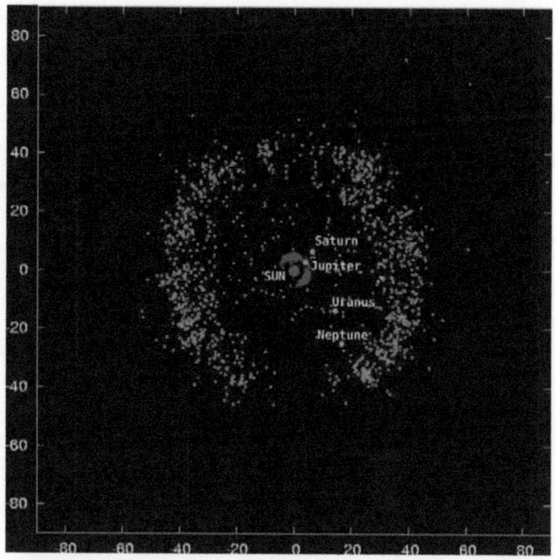

Known objects in the Kuiper belt, derived from data from the Minor Planet Center. Objects in the main belt are colored green, while scattered objects are colored orange. The four outer planets are blue. Neptune's few known trojans are yellow, while Jupiter's are pink. The scattered objects between Jupiter's orbit and the Kuiper belt are known as centaurs. The scale is in astronomical units. The pronounced gap at the bottom is due to difficulties in detection against the background of the plane of the Milky Way.

Since the belt was discovered in 1992,[6] the number of known Kuiper belt objects (KBOs) has increased to over a thousand, and more than 100,000 KBOs over 100 km (62 mi) in diameter are believed to exist.[7] The Kuiper belt was initially thought to be the main repository for periodic comets, those with orbits lasting less than 200 years. However, studies since the mid-1990s have shown that the classical belt is dynamically stable, and that comets' true place of origin is the scattered disc, a dynamically active zone created by the outward motion of Neptune 4.5 billion years ago;[8] scattered disc objects such as Eris have extremely eccentric orbits that take them as far as 100 AU from the Sun.[nb 1]

The Kuiper belt should not be confused with the hypothesized Oort cloud, which is a thousand times more distant. The objects within the Kuiper belt, together with the members of the scattered disc and any potential Hills cloud or Oort cloud objects, are collectively referred to as trans-Neptunian objects (TNOs).[9]

Pluto is the largest known member of the Kuiper belt, and the second largest known TNO, after the scattered-disc object Eris.[nb 1] Originally considered a planet, Pluto's status as part of the Kuiper belt caused it to be reclassified as a "dwarf planet" in 2006. It is compositionally similar to many other objects of the Kuiper belt, and its orbital period is characteristic of a class of KBOs known as "plutinos" which share the same 2:3 resonance with Neptune.

History

Since the discovery of Pluto, many have speculated that it might not be alone. The region now called the Kuiper belt had been hypothesized in various forms for decades. It was only in 1992 that the first direct evidence for its existence was found. The number and variety of prior speculations on the nature of the Kuiper belt have led to continued uncertainty as to who deserves credit for first proposing it.

Hypotheses

The first astronomer to suggest the existence of a trans-Neptunian population was Frederick C. Leonard. In 1930, soon after Pluto's discovery by Clyde Tombaugh, Leonard pondered whether it was "not likely that in Pluto there has come to light the *first* of a *series* of ultra-Neptunian bodies, the remaining members of which still await discovery but which are destined eventually to be detected".[10]

In 1943, in the *Journal of the British Astronomical Association*, Kenneth Edgeworth hypothesized that, in the region beyond Neptune, the material within the primordial solar nebula was too widely spaced to condense into planets, and so rather condensed into a myriad of smaller bodies. From this he concluded that "the outer region of the solar system, beyond the orbits of the planets, is occupied by a very large number of comparatively small bodies"[11] and that, from time to time, one of their number "wanders from its own sphere and appears as an occasional visitor to the inner solar system",[12] becoming a comet.

Astronomer Gerard Kuiper, after whom the Kuiper belt is named

In 1951, in an article for the journal *Astrophysics*, Gerard Kuiper speculated on a similar disc having formed early in the Solar System's evolution; however, he did not believe that such a belt still existed today. Kuiper was operating on the assumption common in his time, that Pluto was the size of the Earth, and had therefore scattered these bodies out toward the Oort cloud or out of the Solar System. Were Kuiper's hypothesis correct, there would not be a Kuiper belt today.[13]

The hypothesis took many other forms in the following decades. In 1962, physicist Al G.W. Cameron postulated the existence of "a tremendous mass of small material on the outskirts of the solar system",[14] while in 1964, Fred Whipple, who popularised the famous "dirty snowball" hypothesis for cometary structure, thought that a "comet belt" might be massive enough to cause the purported discrepancies in the orbit of Uranus that had sparked the search for Planet X, or at the very least, to affect the orbits of known comets.[15] Observation, however, ruled out this hypothesis.[14]

In 1977, Charles Kowal discovered 2060 Chiron, an icy planetoid with an orbit between Saturn and Uranus. He used a blink comparator; the same device that had allowed Clyde Tombaugh to discover Pluto nearly 50 years before.[16] In 1992, another object, 5145 Pholus, was discovered in a similar orbit.[17] Today, an entire population of comet-like bodies, the centaurs, is known to exist in the region between Jupiter and Neptune. The centaurs' orbits are unstable and have dynamical lifetimes of a few million years.[18] From the time of Chiron's discovery, astronomers speculated that they therefore must be frequently replenished by some outer reservoir.[19]

Further evidence for the belt's existence later emerged from the study of comets. That comets have finite lifespans has been known for some time. As they approach the Sun, its heat causes their volatile surfaces to sublimate into space, eating them gradually away. In order to still be visible over the age of the Solar System, they must be frequently replenished.[20] One such area of replenishment is the Oort cloud, the spherical swarm of comets extending beyond 50 000 AU from the Sun first hypothesised by astronomer Jan Oort in 1950.[21] It is believed to be the point of origin for long-period comets, those, like Hale–Bopp, with orbits lasting thousands of years.

There is however another comet population, known as short-period or periodic comets; those, like Halley, with orbits lasting less than 200 years. By the 1970s, the rate at which short-period comets were being discovered was becoming increasingly inconsistent with their having emerged solely from the Oort cloud.[22] For an Oort cloud object to become a short-period comet, it would first have to be captured by the giant planets. In 1980, in the Monthly Notices of the Royal Astronomical Society, Julio Fernandez stated that for every short-period comet to be sent into the inner Solar System from the Oort cloud, 600 would have to be ejected into interstellar space. He speculated that a comet belt from between 35 and 50 AU would be required to account for the observed number of comets.[23] Following up on Fernandez's work, in 1988 the Canadian team of Martin Duncan, Tom Quinn and Scott Tremaine ran a number of computer simulations to determine if all observed comets could have arrived from the Oort cloud. They found that the Oort cloud could not account for all short-period comets, particularly as short-period comets are clustered near the plane of the Solar System, whereas Oort-cloud comets tend to arrive from any point in the sky. With a belt as Fernandez described it added to the formulations, the simulations matched observations.[24] Reportedly because the words "Kuiper" and "comet belt" appeared in the opening sentence of Fernandez's paper, Tremaine named this hypothetical region the "Kuiper belt".[25]

Discovery

The array of telescopes atop Mauna Kea, with which the Kuiper belt was discovered

In 1987, astronomer David Jewitt, then at MIT, became increasingly puzzled by "the apparent emptiness of the outer Solar System".[6] He encouraged then-graduate student Jane Luu to aid him in his endeavour to locate another object beyond Pluto's orbit, because, as he told her, "If we don't, nobody will."[26] Using telescopes at the Kitt Peak National Observatory in Arizona and the Cerro Tololo Inter-American Observatory in Chile, Jewitt and Luu conducted their search in much the same way as Clyde Tombaugh and Charles Kowal had, with a blink comparator.[26] Initially, examination of each pair of plates took about eight hours,[27] but the process was sped up with the arrival of electronic charge-coupled devices or CCDs, which, though their field of view was narrower, were not only more efficient at collecting light (they retained 90 percent of the light that hit them, rather than the ten percent achieved by photographs) but allowed the blinking process to be done virtually, on a computer screen. Today, CCDs form the basis for most astronomical detectors.[28] In 1988, Jewitt moved to the Institute of Astronomy at the University of Hawaii. Luu later joined him to work at the University of Hawaii's 2.24 m telescope at Mauna Kea.[29] Eventually, the field of view for CCDs had increased to 1024 by 1024 pixels, which allowed searches to be conducted far more rapidly.[30] Finally, after five years of searching, on August 30, 1992, Jewitt and Luu announced the "Discovery of the candidate Kuiper belt object" (15760) 1992 QB_1.[6] Six months later, they discovered a second object in the region, (181708) 1993 FW.[31]

Studies since the trans-Neptunian region was first charted have shown that the region now called the Kuiper belt is not the point of origin for short-period comets, but that they instead derive from a linked population called the scattered disc. The scattered disc was created when Neptune migrated outward into the proto-Kuiper belt, which at the time was much closer to the Sun, and left in its wake a population of dynamically stable objects which could never be affected by its orbit (the Kuiper belt proper), and a population whose perihelia are close enough that Neptune can still disturb them as it travels around the Sun (the scattered disc). Because the scattered disc is dynamically active and the Kuiper belt relatively dynamically stable, the scattered disc is now seen as the most likely point of origin for periodic comets.[8]

Name

Astronomers sometimes use the alternative name Edgeworth–Kuiper belt to credit Edgeworth, and KBOs are occasionally referred to as EKOs. However, Brian Marsden claims that neither deserves true credit: "Neither Edgeworth nor Kuiper wrote about anything remotely like what we are now seeing, but Fred Whipple did."[32] Conversely, David Jewitt comments that, "If anything ... Fernandez most nearly deserves the credit for predicting the Kuiper Belt."[13]

KBOs are sometimes called **kuiperoids**, a name suggested by Clyde Tombaugh.[33] The term **trans-Neptunian object** (TNO) is recommended for objects in the belt by several scientific groups because the term is less controversial than all others—it is not a synonym though, as TNOs include all objects orbiting the Sun past the orbit of Neptune, not just those in the Kuiper belt.

Origins

The precise origins of the Kuiper belt and its complex structure are still unclear, and astronomers are awaiting the completion of several wide-field survey telescopes such as Pan-STARRS and the future LSST, which should reveal many currently unknown KBOs. These surveys will provide data that will help determine answers to these questions.[2]

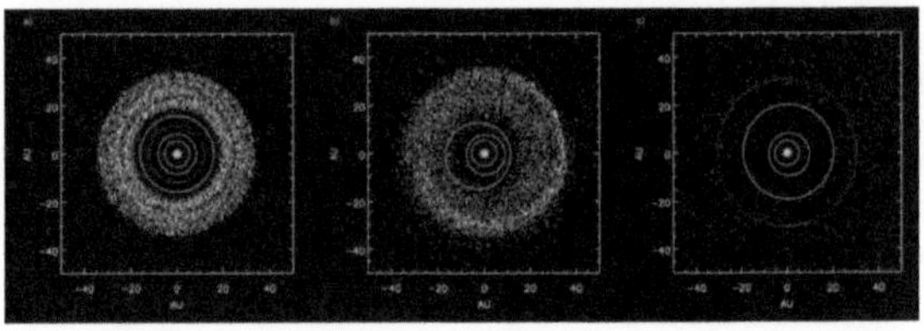

Simulation showing outer planets and Kuiper belt: a) before Jupiter/Saturn 2:1 resonance, b) scattering of Kuiper belt objects into the Solar System after the orbital shift of Neptune, c) after ejection of Kuiper belt bodies by Jupiter

The Kuiper belt is believed to consist of planetesimals; fragments from the original protoplanetary disc around the Sun that failed to fully coalesce into planets and instead formed into smaller bodies, the largest less than 3000 kilometres (mi) in diameter.

Modern computer simulations show the Kuiper belt to have been strongly influenced by Jupiter and Neptune, and also suggest that neither Uranus nor Neptune could have formed in their present positions, as too little primordial matter existed at that range to produce objects of such high mass. Instead, these planets are believed to have formed closer to Jupiter. Scattering of planetesimals early in the Solar System's history would have led to migration of the orbits of the giant planets; Saturn, Uranus and Neptune drifted outwards while Jupiter drifted inwards. Eventually, the orbits shifted to the point where Jupiter and Saturn reached an exact 2:1 resonance; Jupiter orbited the Sun twice for every one Saturn orbit. The gravitational repercussions of such a resonance ultimately disrupted the orbits of Uranus and Neptune, causing Neptune's orbit to become more eccentric and move outward into the primordial planetesimal disk, which sent the disk into temporary chaos.[34] [35] [36] As Neptune's orbit expanded, it excited and scattered many TNO planetesimals into higher and more eccentric orbits.[37] Many more were scattered inward, often to be scattered again and in some cases ejected by Jupiter. The process is thought to have reduced the primordial Kuiper belt population by 99% or more, and to have shifted the distribution of the surviving members outward.[36]

However, this currently most popular model, the "Nice model", still fails to account for some of the characteristics of the distribution and, quoting one of the scientific articles,[38] the problems "continue to challenge analytical techniques and the fastest numerical modeling hardware and software". The model predicts a higher average eccentricity in classical KBO orbits than is observed (0.10–0.13 versus 0.07).[36] The frequency of paired objects, many of which are far apart and loosely bound, also poses a problem for the model.[39]

Structure

At its fullest extent, including its outlying regions, the Kuiper belt stretches from roughly 30 to 55 AU. However, the main body of the belt is generally accepted to extend from the 2:3 resonance (see below) at 39.5 AU to the 1:2 resonance at roughly 48 AU.[40] The Kuiper belt is quite thick, with the main concentration extending as much as ten degrees outside the ecliptic plane and a more diffuse distribution of objects extending several times farther. Overall it more resembles a torus or doughnut than a belt.[41] Its mean position is inclined to the ecliptic by 1.86 degrees.[42]

Dust in the Kuiper belt creates a faint infrared disk.

The presence of Neptune has a profound effect on the Kuiper belt's structure due to orbital resonances. Over a timescale comparable to the age of the Solar System, Neptune's gravity destabilises the orbits of any objects which happen to lie in certain regions, and either sends them into the inner Solar System or out into the scattered disc or interstellar space. This causes the Kuiper belt to possess pronounced gaps in its current layout, similar to the Kirkwood gaps in the asteroid belt. In the region between 40 and 42 AU, for instance, no objects can retain a stable orbit over such times, and any observed in that region must have migrated there relatively recently.[43]

Classical belt

Between the 2:3 and 1:2 resonances with Neptune, at approximately 42–48 AU, the gravitational influence of Neptune is negligible, and objects can exist with their orbits essentially unmolested. This region is known as the classical Kuiper belt, and its members comprise roughly two thirds of KBOs observed to date.[44] [45] Because the first modern KBO discovered, (15760) 1992 QB_1, is considered the prototype of this group, classical KBOs are often referred to as cubewanos ("Q-B-1-os").[46] [47] The guidelines established by the IAU demand that classical KBOs be given names of mythological beings associated with creation.[48]

The classical Kuiper belt appears to be a composite of two separate populations. The first, known as the "dynamically cold" population, has orbits much like the planets; nearly circular, with an orbital eccentricity of less than 0.1, and with relatively low inclinations up to about 10° (they lie close to the plane of the Solar System rather than at an angle). The second, the "dynamically hot" population, has orbits much more inclined to the ecliptic, by up to 30°. The two populations have been named this way not because of any major difference in temperature, but from analogy to particles in a gas, which increase their relative velocity as they become heated up.[49] The two populations not only possess different orbits, but different colors; the cold population is markedly redder than the hot. If this is a reflection of different compositions, it suggests they formed in different regions. The hot population is believed to have formed near Jupiter, and to have been ejected out by movements among the gas giants. The cold population, on the other hand, has been proposed to have formed more or less in its current position, although it might also have been later swept outwards by Neptune during its migration,[2] [50] particularly if Neptune's eccentricity was transiently increased.[36] While the Nice model appears to be able to at least partially explain a compositional difference, it has also been suggested the color difference may reflect differences in surface evolution.[36]

Resonances

When an object's orbital period is an exact ratio of Neptune's (a situation called a mean motion resonance), then it can become locked in a synchronised motion with Neptune and avoid being perturbed away if their relative alignments are appropriate. If, for instance, an object is in just the right kind of orbit so that it orbits the Sun two times for every three Neptune orbits, and if it reaches perihelion with Neptune a quarter of an orbit away from it, then whenever it returns to perihelion, Neptune will always be in about the same relative position as it began, since it will have completed 1½ orbits in the same time. This is known as the 2:3 (or 3:2) resonance, and it corresponds to a characteristic semi-major axis of about 39.4 AU. This 2:3 resonance is populated by about 200 known objects,[51] including Pluto together with its moons. In recognition of this, the members of this family are known as plutinos. Many plutinos, including Pluto, have orbits which cross that of Neptune, though their resonance means they can never collide. Plutinos have high orbital eccentricities, suggesting that they are not native to their current positions but were instead thrown haphazardly into their orbits by the migrating Neptune.[52] IAU guidelines dictate that all plutinos must, like Pluto, be named for underworld deities.[48] The 1:2 resonance (whose objects complete half an orbit for each of Neptune's) corresponds to semi-major axes of ~47.7AU, and is sparsely populated.[53] Its residents are sometimes referred to as twotinos. Other resonances also exist at 3:4, 3:5, 4:7 and 2:5.[54] Neptune possesses a number of trojan objects, which occupy its L_4 and L_5 points; gravitationally stable regions leading and trailing it in its orbit. Neptune trojans are often described as being in a 1:1 resonance with Neptune. Neptune trojans typically have very stable orbits.

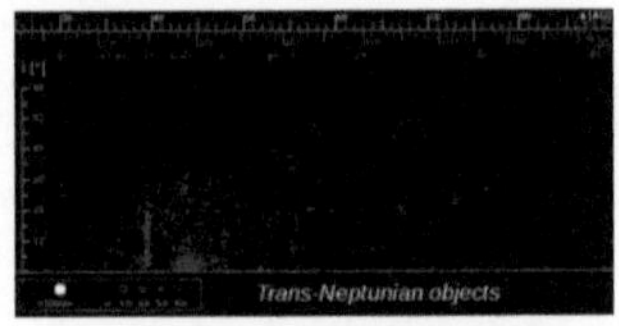

Distribution of cubewanos (blue), Resonant trans-Neptunian objects (red) and near scattered objects (grey).

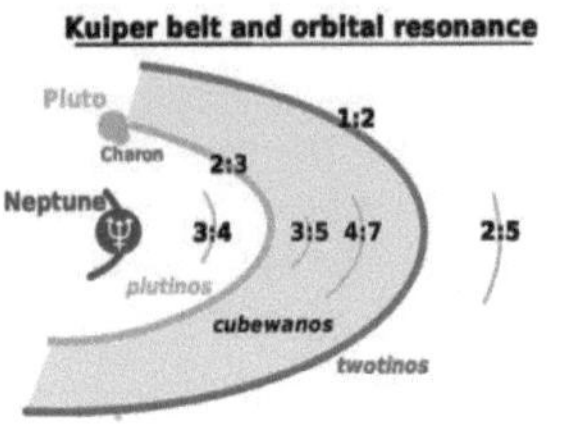

Orbit classification (schematic of semi-major axes)

Additionally, there is a relative absence of objects with semi-major axes below 39 AU which cannot apparently be explained by the present resonances. The currently accepted hypothesis for the cause of this is that as Neptune migrated outward, unstable orbital resonances moved gradually through this region, and thus any objects within it were swept up, or gravitationally ejected from it.[55]

"Kuiper cliff"

The 1:2 resonance appears to be an edge beyond which few objects are known. It is not clear whether it is actually the outer edge of the classical belt or just the beginning of a broad gap. Objects have been detected at the 2:5 resonance at roughly 55 AU, well outside the classical belt; however, predictions of a large number of bodies in classical orbits between these resonances have not been verified through observation.[52]

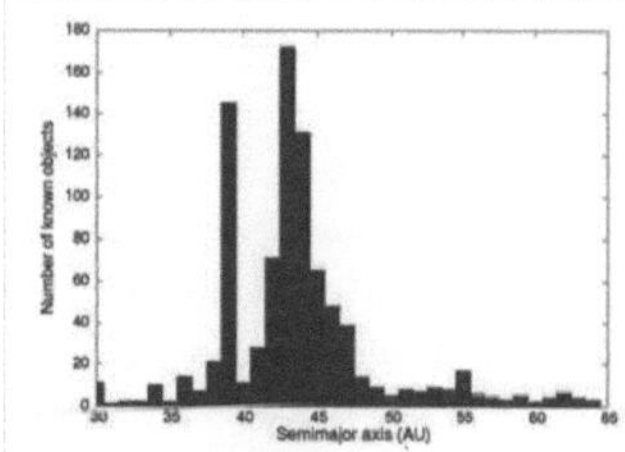

Graph showing the numbers of KBOs for a given distance from the Sun. The plutinos are the "spike" at 39 AU, while the classicals are between 42 and 47 AU, the twotinos are at 48 AU, and the 5:2 resonance is at 55 AU.

Earlier models of the Kuiper belt had suggested that the number of large objects would increase by a factor of two beyond 50 AU,[56] so this sudden drastic falloff, known as the "Kuiper cliff", was completely unexpected, and its cause, to date, is unknown. In 2003, Bernstein and Trilling *et al.* found evidence that the rapid decline in objects of 100 km or more in radius beyond 50 AU is real, and not due to observational bias. Possible explanations include that material at that distance is too scarce or too scattered to accrete into large objects, or that subsequent processes removed or destroyed those which did form.[57] Patryk Lykawka of Kobe University has claimed that the gravitational attraction of an unseen large planetary object, perhaps the size of Earth or Mars, might be responsible.[58] [59]

Composition

Studies of the Kuiper belt since its discovery have generally indicated that its members are primarily composed of ices: a mixture of light hydrocarbons (such as methane), ammonia, and water ice,[60] a composition they share with comets.[61] The low densities observed in those KBOs whose diameter is known, (less than 1 g cm^{-3}) is consistent with an icy makeup.[60] The temperature of the belt is only about 50K,[62] so many compounds that would be gaseous closer to the Sun remain solid.

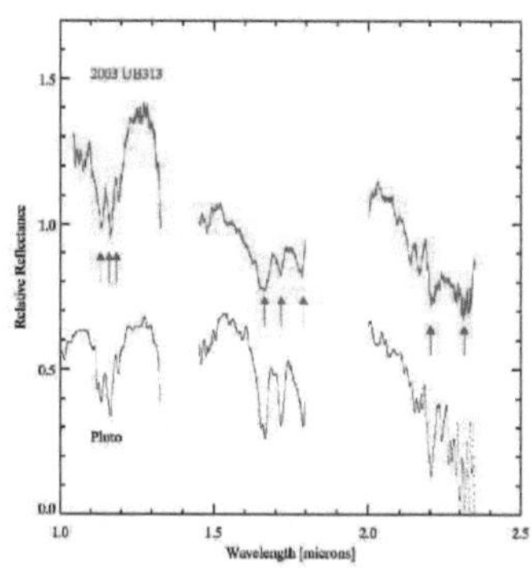

The infrared spectra of both Eris and Pluto, highlighting their common methane absorption lines

Due to their small size and extreme distance from Earth, the chemical makeup of KBOs is very difficult to determine. The principal method by which astronomers determine the composition of a celestial object is spectroscopy. When an object's light is broken into its component colors, an image akin to a rainbow is formed. This image is called a spectrum. Different substances absorb light at different wavelengths, and when the spectrum for a specific object is unravelled, dark lines (called absorption lines) appear where the substances within it have absorbed that particular wavelength of light. Every element or compound has its own unique spectroscopic signature, and by reading an object's full spectral "fingerprint", astronomers can determine what it is made of.

Initially, such detailed analysis of KBOs was impossible, and so astronomers were only able to determine the most basic facts about their makeup, primarily their color.[63] These first data showed a broad range of colors among KBOs, ranging from neutral grey to deep red.[64] This suggested that their surfaces were composed of a wide range of compounds, from dirty ices to hydrocarbons.[64] This diversity was startling, as astronomers had expected KBOs to be uniformly dark, having lost most of their volatile ices to the effects of cosmic rays.[65] Various solutions were suggested for this discrepancy, including resurfacing by impacts or outgassing.[63] However, Jewitt and Luu's spectral analysis of the known Kuiper belt objects in 2001 found that the variation in color was too extreme to be easily explained by random impacts.[66]

Although to date most KBOs still appear spectrally featureless due to their faintness, there have been a number of successes in determining their composition.[62] In 1996, Robert H. Brown *et al.* obtained spectroscopic data on the KBO 1993 SC, revealing its surface composition to be markedly similar to that of Pluto, as well as Neptune's moon Triton, possessing large amounts of methane ice.[67]

Water ice has been detected in several KBOs, including 1996 TO_{66},[68] 38628 Huya and 20000 Varuna.[69] In 2004, Mike Brown *et al.* determined the existence of crystalline water ice and ammonia hydrate on one of the largest known KBOs, 50000 Quaoar. Both of these substances would have been destroyed over the age of the Solar System, suggesting that Quaoar had been recently resurfaced, either by internal tectonic activity or by meteorite impacts.[62]

Mass and size distribution

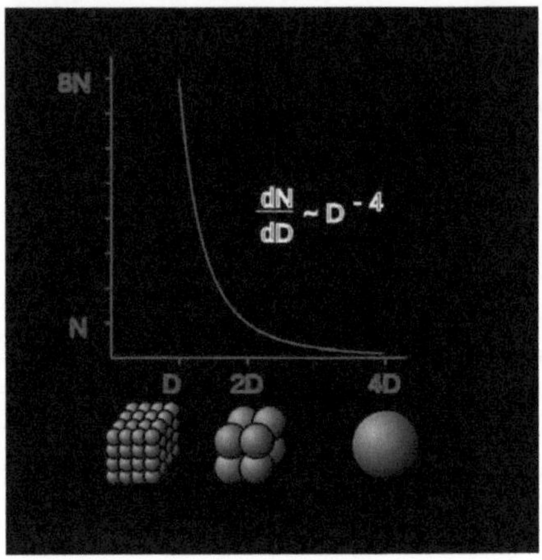

Illustration of the power law.

Despite its vast extent, the collective mass of the Kuiper belt is relatively low. The total mass is estimated to range between a 25th and 10th the mass of the Earth[70] with some estimates placing it at a thirtieth an Earth mass.[71] Conversely, models of the Solar System's formation predict a collective mass for the Kuiper belt of 30 Earth masses.[2] This missing >99% of the mass can hardly be dismissed, as it is required for the accretion of any KBOs larger than 100 km (62 mi) in diameter. If the Kuiper belt had always had its current low density these large objects simply could not have formed.[2] Moreover, the eccentricity and inclination of current orbits makes the encounters quite "violent" resulting in destruction rather than accretion. It appears that either the current residents of the Kuiper belt have been created closer to the Sun or some mechanism dispersed the original mass. Neptune's current influence is too weak to explain such a massive "vacuuming", though the Nice model proposes that it could have been the cause of mass removal in the past. While the question remains open, the conjectures vary from a passing star scenario to grinding of smaller objects, via collisions, into dust small enough to be affected by solar radiation.[50]

Bright objects are rare compared with the dominant dim population, as expected from accretion models of origin, given that only some objects of a given size would have grown further. This relationship N(D), the population expressed as a function of the diameter, referred to as brightness slope, has been confirmed by observations. The slope is inversely proportional to some power of the diameter D.

where the current measures[72] give $q = 4 \pm 0.5$.

Less formally, there are for instance 8 ($=2^3$) times more objects in 100–200 km range than objects in 200–400 km range. In other words, for every object with the diameter of 1000 km (621 mi) there should be around 1000 ($=10^3$) objects with diameter of 100 km (62 mi).

The law is expressed in this differential form rather than as a cumulative cubic relationship, because only the middle part of the slope can be measured; the law must break at smaller sizes, beyond the current measure.

Of course, only the magnitude is actually known, the size is inferred assuming albedo (not a safe assumption for larger objects).

Since January 2010, the smallest Kuiper belt object discovered to date spans 980 m across.[73]

Scattered objects

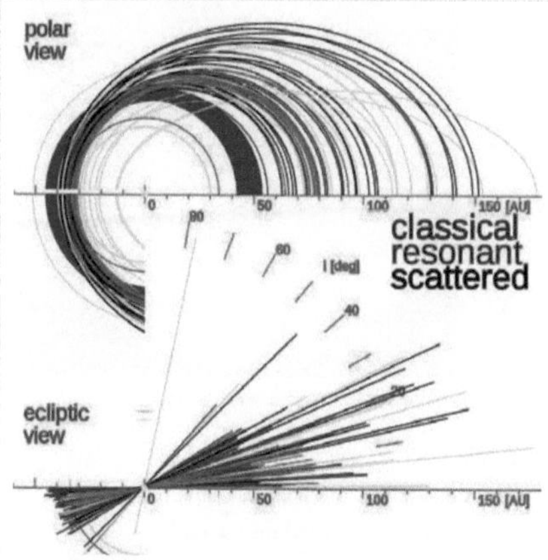

Comparison of the orbits of scattered disc objects (black), classical KBOs (blue), and 2:5 resonant objects (green). Orbits of other KBOs are gray.

The scattered disc is a sparsely populated region, overlapping with the Kuiper belt but extending as far as 100 AU and farther. Scattered disc objects (SDOs) travel in highly elliptical orbits, usually also highly inclined to the ecliptic. Most models of Solar System formation show both KBOs and SDOs first forming in a primordial comet belt, while later gravitational interactions, particularly with Neptune, sent the objects spiraling outward; some into stable orbits (the KBOs) and some into unstable orbits, becoming the scattered disc.[8] Due to its unstable nature, the scattered disc is believed to be the point of origin for many of the Solar System's short-period comets. Their dynamic orbits occasionally force them into the inner Solar System, becoming first centaurs, and then short-period comets.[8]

According to the Minor Planet Center, which officially catalogues all trans-Neptunian objects, a KBO, strictly speaking, is any object that orbits exclusively within the defined Kuiper belt region regardless of origin or composition. Objects found outside the belt are classed as scattered objects.[74] However, in some scientific circles the term "Kuiper belt object" has become synonymous with any icy minor planet native to the outer Solar System believed to have been part of that initial class, even if its orbit during the bulk of Solar System history has been beyond the Kuiper belt (e.g. in the scattered-disc region). They often describe scattered disc objects as "scattered Kuiper belt objects".[75] Eris, which is known to be more massive than Pluto, is often referred to as a KBO, but is technically an SDO.[74] A consensus among astronomers as to the precise definition of the Kuiper belt has yet to be reached, and this issue remains unresolved.

The centaurs, which are not normally considered part of the Kuiper belt, are also believed to be scattered objects, the only difference being that they were scattered inward, rather than outward. The Minor Planet Center groups the centaurs and the SDOs together as scattered objects.[74]

Triton

Neptune's moon Triton

During its period of migration, Neptune is thought to have captured one of the larger KBOs and set it in orbit around itself. This is its moon Triton, which is the only large moon in the Solar System to have a retrograde orbit; it orbits in the opposite direction to Neptune's rotation. This suggests that, unlike the large moons of Jupiter and Saturn, which are thought to have coalesced from spinning discs of material encircling their young parent planets, Triton was a fully formed body that was captured from surrounding space. Gravitational capture of an object is not easy; it requires some mechanism to slow the object down enough to be snared by the larger object's gravity. Triton may have encountered Neptune as part of a binary (many KBOs are members of binaries; see below); ejection of the other member of the binary by Neptune could then explain Triton's capture.[76] Triton is only slightly larger than Pluto, and spectral analysis of both worlds shows that they are largely composed of similar materials, such as methane and carbon monoxide. All this points to the conclusion that Triton was once a KBO that was captured by Neptune during its outward migration.[77]

Largest KBOs

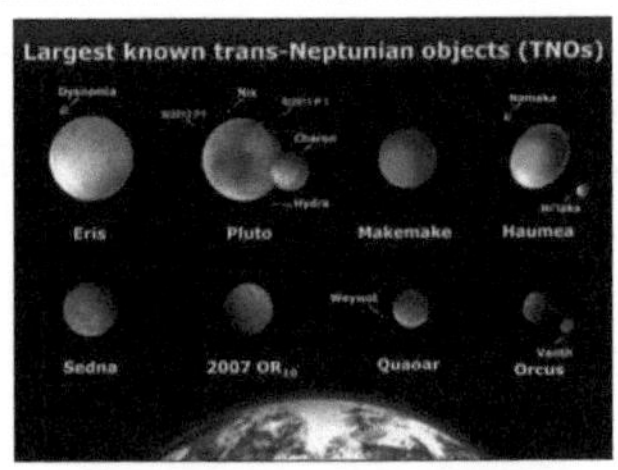

10}}, Quaoar, Orcus, and Earth. These eight trans-Neptunian objects have the brightest absolute magnitudes, although several other TNOs have been found to be physically larger than Orcus, and several more may yet be found.

Since the year 2000, a number of KBOs with diameters of between 500 and 1500 km (932 mi), more than half that of Pluto, have been discovered. 50000 Quaoar, a classical KBO discovered in 2002, is over 1,200 km across. Makemake (originally (136472) 2005 FY_9, nicknamed "Easterbunny") and Haumea (originally (136108) 2003 EL_{61}, nicknamed "Santa"), both announced on July 29, 2005, are larger still. Other objects, such as 28978 Ixion (discovered in 2001) and 20000 Varuna (discovered in 2000) measure roughly 500 km (311 mi) across.[2]

Pluto

The discovery of these large KBOs in similar orbits to Pluto led many to conclude that, bar its relative size, Pluto was not particularly different from other members of the Kuiper belt. Not only did these objects approach Pluto in size, but many also possessed satellites, and were of similar composition (methane and carbon monoxide have been found both on Pluto and on the largest KBOs).[2] Thus, just as Ceres was considered a planet before the discovery of its fellow asteroids, some began to suggest that Pluto might also be reclassified.

The issue was brought to a head by the discovery of Eris, an object in the scattered disc far beyond the Kuiper belt, that is now known to be 27 percent more massive than Pluto.[78] In response, the International Astronomical Union (IAU), was forced to define what a planet is for the first time, and in so doing included in their definition that a planet must have "cleared the neighbourhood around its orbit".[79] As Pluto shared its orbit with so many KBOs, it was deemed not to have cleared its orbit, and was thus reclassified from a planet to a member of the Kuiper belt.

Although Pluto is currently the largest KBO, there are two known larger objects currently outside the Kuiper belt which probably originated in the Kuiper belt. These are Eris and Neptune's moon Triton (which, as explained above, is probably a captured KBO).

As of 2008, only five objects in the Solar System, Ceres, Eris, and the KBOs Pluto, Makemake and Haumea, are listed as dwarf planets by the IAU. However, 90482 Orcus, 28978 Ixion and many other Kuiper-belt objects are large enough to be in hydrostatic equilibrium; most of them will probably qualify when more is known about them.[80] [81] [82]

Satellites

Of the four largest TNOs, three (Eris, Pluto, and Haumea) possess satellites, and two have more than one. A higher percentage of the larger KBOs possess satellites than the smaller objects in the Kuiper belt, suggesting that a different formation mechanism was responsible.[83] There are also a high number of binaries (two objects close enough in mass to be orbiting "each other") in the Kuiper belt. The most notable example is the Pluto–Charon binary, but it is estimated that around 11 percent of KBOs exist in binaries.[84]

Exploration

Artist's conception of *New Horizons* at Pluto

On January 19, 2006, the first spacecraft mission to explore the Kuiper belt, *New Horizons*, was launched. The mission, headed by Alan Stern of the Southwest Research Institute, will arrive at Pluto on July 14, 2015, and, circumstances permitting, will continue on to study another as-yet undetermined KBO. Any KBO chosen will be between 25 and 55 miles (40 to 90 km) in diameter and, ideally, white or grey, to contrast with Pluto's reddish color.[85] John Spencer, an astronomer on the *New Horizons* mission team, says that no target for a post-Pluto Kuiper belt encounter has yet been selected, as they are awaiting data from the Pan-STARRS survey project to ensure as wide a field of options as possible.[86] The Pan-STARRS project, partially operational since May 2010,[87] will, when fully online, survey the entire sky with four 1.4 gigapixel digital cameras to detect any moving objects, from near-earth objects to KBOs.[88] To speed up the detection process, the New Horizons team established Ice Hunters, a citizen science project that allowed members of the public to participate in the search for suitable KBO targets;[89] [90] [91] the project has subsequently been transferred to another site, Ice Investigators,[92] produced by CosmoQuest.[93]

Other Kuiper belts

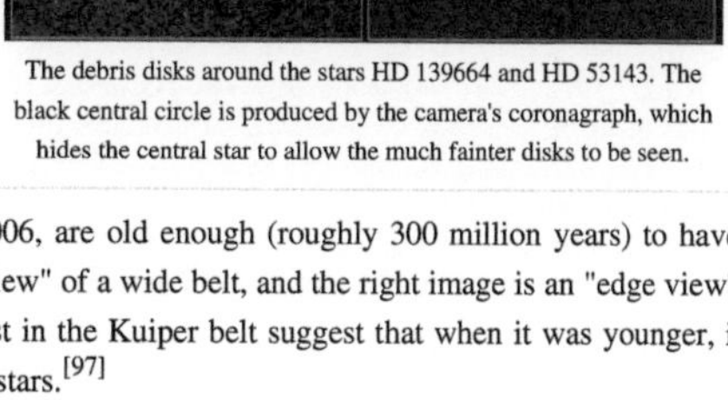

The debris disks around the stars HD 139664 and HD 53143. The black central circle is produced by the camera's coronagraph, which hides the central star to allow the much fainter disks to be seen.

By 2006, astronomers had resolved dust disks believed to be Kuiper belt-like structures around nine stars other than the Sun. They appear to fall into two categories: wide belts, with radii of over 50 AU, and narrow belts (like our own Kuiper belt) with radii of between 20 and 30 AU and relatively sharp boundaries.[94] Beyond this, 15–20% of solar-type stars have an observed infrared excess which is believed to indicate massive Kuiper belt-like structures.[95] Most known debris discs around other stars are fairly young, but the two images on the right, taken by the Hubble Space Telescope in January 2006, are old enough (roughly 300 million years) to have settled into stable configurations. The left image is a "top view" of a wide belt, and the right image is an "edge view" of a narrow belt.[94] [96] Supercomputer simulations of dust in the Kuiper belt suggest that when it was younger, it may have resembled the narrow rings seen around younger stars.[97]

See also

- List of possible dwarf planets
- List of trans-Neptunian objects

Notes

The literature is inconsistent in the usage of the terms "scattered disc" and "Kuiper belt". For some, they are distinct populations; for others, the scattered disc is part of the Kuiper belt. Authors may even switch between these two uses in a single publication.[98] Because the International Astronomical Union's Minor Planet Center, the body responsible for cataloguing minor planets in the Solar System, makes the distinction,[99] the current editorial choice for Wikipedia articles on the trans-Neptunian region is to make this distinction as well. This choice means that, on Wikipedia, Eris, the largest known trans-Neptunian object, is not part of the Kuiper belt, and so Pluto becomes the largest Kuiper belt object.

References

[1] Alan Stern; Colwell, Joshua E. (1997). "Collisional Erosion in the Primordial Edgeworth-Kuiper Belt and the Generation of the 30–50 AU Kuiper Gap". *The Astrophysical Journal* **490** (2): 879–882. Bibcode 1997ApJ...490..879S. doi:10.1086/304912.

[2] Audrey Delsanti and David Jewitt. "The Solar System Beyond The Planets" (http://web.archive.org/web/20070925203400/http://www.ifa.hawaii.edu/faculty/jewitt/papers/2006/DJ06.pdf). *Institute for Astronomy, University of Hawaii*. Archived from the original (http://www2.ess.ucla.edu/~jewitt/papers/2006/DJ06.pdf) on September 25, 2007. . Retrieved March 9, 2007.

[3] Krasinsky, G. A.; Pitjeva, E. V.; Vasilyev, M. V.; Yagudina, E. I. (July 2002). "Hidden Mass in the Asteroid Belt". *Icarus* **158** (1): 98–105. Bibcode 2002Icar..158...98K. doi:10.1006/icar.2002.6837.

[4] Johnson, Torrence V.; and Lunine, Jonathan I.; *Saturn's moon Phoebe as a captured body from the outer Solar System*, Nature, Vol. 435, pp. 69–71

[5] Craig B. Agnor & Douglas P. Hamilton (2006). "Neptune's capture of its moon Triton in a binary-planet gravitational encounter" (http://web.archive.org/web/20070621182809/http://www.es.ucsc.edu/~cagnor/papers_pdf/2006AgnorHamilton.pdf). *Nature*. Archived from the original (http://www.es.ucsc.edu/~cagnor/papers_pdf/2006AgnorHamilton.pdf) on June 21, 2007. . Retrieved June 20, 2006.

[6] Jewitt, David; Luu, Jane (1993). "Discovery of the candidate Kuiper belt object 1992 QB1". *Nature* **362** (6422): 730. Bibcode 1993Natur.362..730J. doi:10.1038/362730a0.

[7] NEW HORIZONS *The PI's Perspective* (http://pluto.jhuapl.edu/overview/piPerspective.php?page=piPerspective_08_24_2012)

[8] Harold F. Levison, Luke Donnes (2007). "Comet Populations and Cometary Dynamics". In Lucy Ann Adams McFadden, Paul Robert Weissman, Torrence V. Johnson. *Encyclopedia of the Solar System* (2nd ed.). Amsterdam; Boston: Academic Press. pp. 575–588. ISBN 0-12-088589-1.

[9] Gérard FAURE (2004). "Description of the System of Asteroids as of May 20, 2004" (http://web.archive.org/web/20070529003558/http://www.astrosurf.com/aude/map/us/AstFamilies2004-05-20.htm). Archived from the original (http://www.astrosurf.com/aude/map/us/AstFamilies2004-05-20.htm) on May 29, 2007. . Retrieved June 1, 2007.

[10] "What is improper about the term "Kuiper belt"? (or, Why name a thing after a man who didn't believe its existence?)" (http://www.icq.eps.harvard.edu/kb.html). *International Comet Quarterly*. . Retrieved October 24, 2010.

[11] John Davies (2001). *Beyond Pluto: Exploring the outer limits of the solar system*. Cambridge University Press. xii.

[12] Davies, p. 2

[13] David Jewitt. "WHY "KUIPER" BELT?" (http://www2.ess.ucla.edu/~jewitt/kb/gerard.html). *University of Hawaii*. . Retrieved June 14, 2007.

[14] Davies, p. 14

[15] Rao, M. M. (1964). "Decomposition of Vector Measures" (http://www.pnas.org/cgi/reprint/51/5/711.pdf). *Proceedings of the National Academy of Sciences* **51** (5): 771. Bibcode 1964PNAS...51..771R. doi:10.1073/pnas.51.5.771. .

[16] CT Kowal, W Liller, BG Marsden; Liller; Marsden (1977). "The discovery and orbit of /2060/ Chiron". *Hale Observatories, Harvard–Smithsonian Center for Astrophysics* **81**: 245. Bibcode 1979IAUS...81..245K.

[17] JV Scotti, DL Rabinowitz, CS Shoemaker, EM Shoemaker, DH Levy, TM King, EF Helin, J Alu, K Lawrence, RH McNaught, L Frederick, D Tholen, BEA Mueller; Rabinowitz; Shoemaker; Shoemaker; Levy; King; Helin; Alu et al. (1992). "1992 AD". *IAU Circ.* **5434**: 1. Bibcode 1992IAUC.5434....1S.

[18] Horner, J.; Evans, N.W.; Bailey, M. E. (2004). "Simulations of the Population of Centaurs I: The Bulk Statistics". *MNRAS* **354** (3): 798–810. arXiv:astro-ph/0407400. Bibcode 2004MNRAS.354..798H. doi:10.1111/j.1365-2966.2004.08240.x.

[19] Davies p. 38

[20] David Jewitt (2002). "From Kuiper Belt Object to Cometary Nucleus: The Missing Ultrared Matter". *The Astronomical Journal* **123** (2): 1039–1049. Bibcode 2002AJ....123.1039J. doi:10.1086/338692.

[21] Oort, J. H. (1950). "The structure of the cloud of comets surrounding the Solar System and a hypothesis concerning its origin". *Bull. Astron. Inst. Neth.* **11**: 91. Bibcode 1950BAN....11...91O.

[22] Davies p. 39

[23] JA Fernandez (1980). "On the existence of a comet belt beyond Neptune". *Monthly Notices of the Royal Astronomical Society* **192**: 481. Bibcode 1980MNRAS.192..481F.

[24] M. Duncan, T. Quinn, and S. Tremaine (1988). "The origin of short-period comets". *The Astrophysical Journal* **328**: L69. Bibcode 1988ApJ...328L..69D. doi:10.1086/185162.

[25] Davies p. 191

[26] Davies p. 50

[27] Davies p. 51

[28] Davies pp. 52, 54, 56

[29] Davies pp. 57, 62

[30] Davies p. 65

[31] BS Marsden; Jewitt; Marsden (1993). "1993 FW". *Minor Planet Center* **5730**: 1. Bibcode 1993IAUC.5730....1L.

[32] Davies p. 199

[33] Clyde Tombaugh, "The Last Word", Letters to the Editor, *Sky & Telescope*, December, 1994, p. 8

[34] Hansen, K. (7 June 2005). "Orbital shuffle for early solar system" (http://www.geotimes.org/june05/WebExtra060705.html). *Geotimes*. . Retrieved 2007-08-26.

[35] Tsiganis, K.; Gomes, R.; Morbidelli, A.; Levison, H. F. (2005). "Origin of the orbital architecture of the giant planets of the Solar System". *Nature* **435** (7041): 459–461. Bibcode 2005Natur.435..459T. doi:10.1038/nature03539. PMID 15917800.

[36] Levison, H. F.; Morbidelli, A.; Van Laerhoven, C.; Gomes, R. (2008). "Origin of the structure of the Kuiper belt during a dynamical instability in the orbits of Uranus and Neptune". *Icarus* **196** (1): 258–273. arXiv:0712.0553. Bibcode 2008Icar..196..258L. doi:10.1016/j.icarus.2007.11.035.

[37] Thommes, E. W.; Duncan, M. J.; Levison, H. F. (2002). "The Formation of Uranus and Neptune among Jupiter and Saturn". *The Astronomical Journal* **123** (5): 2862. arXiv:astro-ph/0111290. Bibcode 2002AJ....123.2862T. doi:10.1086/339975.

[38] Malhotra, R. (1994). "Nonlinear Resonances in the Solar System". *Physica D* **77**: 289. arXiv:chao-dyn/9406004. Bibcode 1994PhyD...77..289M. doi:10.1016/0167-2789(94)90141-4.

[39] Lovett, R. (2010). "Kuiper Belt may be born of collisions". *Nature*. doi:10.1038/news.2010.522.

[40] M. C. De Sanctis, M. T. Capria, and A. Coradini (2001). "Thermal Evolution and Differentiation of Edgeworth-Kuiper Belt Objects". *The Astronomical Journal* **121** (5): 2792–2799. Bibcode 2001AJ....121.2792D. doi:10.1086/320385.

[41] "Discovering the Edge of the Solar System" (http://web.archive.org/web/20080305191925/http://www.americanscientist.org/template/AssetDetail/assetid/25723/page/2;jsessionid=aaa5LVF0). *American Scientists.org*. 2003. Archived from the original (http://www.americanscientist.org/template/AssetDetail/assetid/25723/page/2;jsessionid=aaa5LVF0) on March 5, 2008. . Retrieved June 23, 2007.

[42] Michael E. Brown, Margaret Pan (2004). "The Plane of the Kuiper Belt". *The Astronomical Journal* **127** (4): 2418–2423. Bibcode 2004AJ....127.2418B. doi:10.1086/382515.

[43] Jean-Marc Petit, Alessandro Morbidelli, Giovanni B. Valsecchi (1998). "Large Scattered Planetesimals and the Excitation of the Small Body Belts" (http://www.obs-nice.fr/morby/papers/6166a.pdf). . Retrieved June 23, 2007.

[44] Lunine, J. (2003). "The Kuiper Belt" (http://www.gsmt.noao.edu/gsmt_swg/SWG_Apr03/The_Kuiper_Belt.pdf). . Retrieved 2007-06-23.

[45] Jewitt, D. (February 2000). "Classical Kuiper Belt Objects (CKBOs)" (http://web.archive.org/web/20070609094740/http://www.ifa.hawaii.edu/~jewitt/kb/kb-classical.html). Archived from the original (http://www2.ess.ucla.edu/~jewitt/kb/kb-classical.html) on 2007-06-09. . Retrieved 2007-06-23.

[46] Murdin, P. (2000). "Cubewano". *The Encyclopedia of Astronomy and Astrophysics*. Bibcode 2000eaa..bookE5403.. doi:10.1888/0333750888/5403. ISBN 0-333-75088-8.

[47] Elliot, J. L.; *et al.* (2005). "The Deep Ecliptic Survey: A Search for Kuiper Belt Objects and Centaurs. II. Dynamical Classification, the Kuiper Belt Plane, and the Core Population" (http://occult.mit.edu/_assets/documents/publications/Elliot2005AJ129.1117.pdf). *The Astronomical Journal* **129**: 1117–1162. Bibcode 2005AJ....129.1117E. doi:10.1086/427395. .

[48] "Naming of Astronomical Objects: Minor Planets" (http://www.iau.org/public_press/themes/naming/#minorplanets). International Astronomical Union. . Retrieved 2008-11-17.

[49] Levison, H. F; Morbidelli, A. (2003). "The formation of the Kuiper belt by the outward transport of bodies during Neptune's migration". *Nature* **426** (6965): 419–421. doi:10.1038/nature02120. PMID 14647375.

[50] Morbidelli, A. (2005). "Origin and Dynamical Evolution of Comets and their Reservoirs". arXiv:astro-ph/0512256 [astro-ph].

[51] "List Of Transneptunian Objects" (http://www.minorplanetcenter.org/iau/lists/TNOs.html). *Minor Planet Center*. . Retrieved June 23, 2007.

[52] Chiang et al. (2003). "Resonance Occupation in the Kuiper Belt: Case Examples of the 5:2 and Trojan Resonances". *The Astronomical Journal* **126** (1): 430–443. arXiv:astro-ph/0301458. Bibcode 2003AJ....126..430C. doi:10.1086/375207.

[53] Wm. Robert Johnston (2007). "Trans-Neptunian Objects" (http://www.johnstonsarchive.net/astro/tnos.html). . Retrieved June 23, 2007.

[54] Davies p. 104

[55] Davies p. 107

[56] E. I. Chiang and M. E. Brown (1999). "Keck Pencil-Beam Survey For Faint Kuiper Belt Objects" (http://www.gps.caltech.edu/~mbrown/papers/ps/kbodeep.pdf). . Retrieved July 1, 2007.

[57] G.M. Bernstein, D.E. Trilling, R.L. Allen, M.E. Brown, M. Holman and R. Malhotra (2004). "The Size Distribution of Trans-Neptunian Bodies" (http://www.gps.caltech.edu/~mbrown/papers/ps/bernstein.pdf). *The Astrophysical Journal* **128** (3): 1364. arXiv:astro-ph/0308467. Bibcode 2004AJ....128.1364B. doi:10.1086/422919. .

[58] Michael Brooks (2007). "13 Things that do not make sense" (http://space.newscientist.com/article.ns?id=mg18524911.600). *NewScientistSpace.com*. . Retrieved June 23, 2007.

[59] Govert Schilling (2008). "The mystery of Planet X" (http://space.newscientist.com/article/mg19726381.600-the-mystery-of-planet-x.html). *New Scientist*. . Retrieved February 8, 2008.

[60] Stephen C. Tegler (2007). "Kuiper Belt Objects: Physical Studies". In Lucy-Ann McFadden *et al.*. *Encyclopedia of the Solar System*. pp. 605–620.

[61] Altwegg, K.; Balsiger, H.; Geiss, J. (1999). *Space Science Reviews* **90**: 3. Bibcode 1999SSRv...90....3A. doi:10.1023/A:1005256607402.

[62] David C. Jewitt & Jane Luu (2004). "Crystalline water ice on the Kuiper belt object (50000) Quaoar" (http://web.archive.org/web/20070621182808/http://www.ifa.hawaii.edu/~jewitt/papers/50000/Quaoar.pdf). Archived from the original (http://www2.ess.ucla.edu/~jewitt/papers/50000/Quaoar.pdf) on June 21, 2007. . Retrieved June 21, 2007.

[63] Dave Jewitt (2004). "Surfaces of Kuiper Belt Objects" (http://web.archive.org/web/20070609094911/http://www.ifa.hawaii.edu/~jewitt/kb/kb-colors.html). *University of Hawaii*. Archived from the original (http://www2.ess.ucla.edu/~jewitt/kb/kb-colors.html) on June 9, 2007. . Retrieved June 21, 2007.

[64] Jewitt, David; Luu, Jane (1998). "Optical-Infrared Spectral Diversity in the Kuiper Belt". *The Astronomical Journal* **115** (4): 1667. Bibcode 1998AJ....115.1667J. doi:10.1086/300299.

[65] Davies p. 118

[66] Jewitt, David C.; Luu, Jane X. (2001). "Colors and Spectra of Kuiper Belt Objects". *The Astronomical Journal* **122** (4): 2099. arXiv:astro-ph/0107277. Bibcode 2001AJ....122.2099J. doi:10.1086/323304.

[67] Brown, R. H.; Cruikshank, DP; Pendleton, Y; Veeder, GJ (1997). "Surface Composition of Kuiper Belt Object 1993SC". *Science* **276** (5314): 937–9. Bibcode 1997Sci...276..937B. doi:10.1126/science.276.5314.937. PMID 9163038.

[68] Brown, Michael E.; Blake, Geoffrey A.; Kessler, Jacqueline E. (2000). "Near-Infrared Spectroscopy of the Bright Kuiper Belt Object 2000 EB173". *The Astrophysical Journal* **543** (2): L163. Bibcode 2000ApJ...543L.163B. doi:10.1086/317277.

[69] Licandro; Oliva; Di MArtino (2001). "NICS-TNG infrared spectroscopy of trans-neptunian objects 2000 EB173 and 2000 WR106". *Astronomy and Astrophysics* **373** (3): L29. arXiv:astro-ph/0105434. Bibcode 2001A&A...373L..29L. doi:10.1051/0004-6361:20010758.

[70] Gladman, Brett; et al (05/April/2001). "The structure of the Kuiper belt". *Astronomical Journal* **122** (2): 1051–1066. Bibcode 2001AJ....122.1051G. doi:10.1086/322080.

[71] Lorenzo Iorio (2007). "Dynamical determination of the mass of the Kuiper Belt from motions of the inner planets of the Solar system". *Monthly Notices of the Royal Astronomical Society* **375** (4): 1311–1314. Bibcode 2007MNRAS.tmp...24I. doi:10.1111/j.1365-2966.2006.11384.x.

[72] Bernstein, G. M.; Trilling, D. E.; Allen, R. L.; Brown, K. E.; Holman, M.; Malhotra, R. (2004). "The size distribution of transneptunian bodies". *The Astronomical Journal* **128** (3): 1364–1390. arXiv:astro-ph/0308467. Bibcode 2004AJ....128.1364B. doi:10.1086/422919.

[73] I, Eric (January 13, 2010). "Hubble Finds Smallest Kuiper Belt Object" (http://www.meteoritesusa.com/meteorite-photos/hubble-finds-smallest-kuiper-belt-object/). *MeteoritesUSA*. . Retrieved January 16, 2011.

[74] "List Of Centaurs and Scattered-Disk Objects" (http://www.minorplanetcenter.org/iau/lists/Centaurs.html). *IAU: Minor Planet Center*. . Retrieved October 27, 2010.

[75] David Jewitt (2005). "The 1000 km Scale KBOs" (http://www2.ess.ucla.edu/~jewitt/kb/big_kbo.html). *University of Hawaii*. . Retrieved July 16, 2006.

[76] Craig B. Agnor & Douglas P. Hamilton (2006). "Neptune's capture of its moon Triton in a binary-planet gravitational encounter" (http://web.archive.org/web/20070621182809/http://www.es.ucsc.edu/~cagnor/papers_pdf/2006AgnorHamilton.pdf). *Nature*. Archived from the original (http://www.es.ucsc.edu/~cagnor/papers_pdf/2006AgnorHamilton.pdf) on June 21, 2007. . Retrieved October 29, 2007.

[77] DALE P. CRUIKSHANK (2004). *TRITON, PLUTO, CENTAURS, AND TRANS-NEPTUNIAN BODIES* (http://books.google.com/?id=MbmiTd3x1UcC&pg=PA421&dq=.+TRITON,+PLUTO,+CENTAURS,+AND+TRANS-NEPTUNIAN+BODIES). Springer. ISBN 978-1-4020-3362-9. . Retrieved June 23, 2007.

[78] Mike Brown (2007). "Dysnomia, the moon of Eris" (http://www.gps.caltech.edu/~mbrown/planetlila/moon/index.html). *CalTech*. . Retrieved June 14, 2007.

[79] "Resolution B5 and B6" (http://www.iau.org/static/resolutions/Resolution_GA26-5-6.pdf). International Astronomical Union. 2006. .

[80] "Ixion" (http://ixion.eightplanets.net/). *eightplanets.net*. . Retrieved June 23, 2007.

[81] John Stansberry; Will Grundy; Mike Brown; Dale Cruikshank; John Spencer; David Trilling; Jean-Luc Margot (2007). "Physical Properties of Kuiper Belt and Centaur Objects: Constraints from Spitzer Space Telescope". arXiv:astro-ph/0702538 [astro-ph].

[82] "IAU Draft Definition of Planet" (http://www.iau.org/iau0601.424.0.html). *IAU*. 2006. . Retrieved October 26, 2007.

[83] Error: Bad DOI specified!

[84] Agnor, C.B.; Hamilton, D.P. (2006). "Neptune's capture of its moon Triton in a binary-planet gravitational encounter" (http://www.astro.umd.edu/~hamilton/research/reprints/AgHam06.pdf). *Nature* **441** (7090): 192–4. Bibcode 2006Natur.441..192A.

doi:10.1038/nature04792. PMID 16688170. .

[85] "New Horizons mission timeline" (http://pluto.jhuapl.edu/mission/mission_timeline.php). *NASA*. . Retrieved March 19, 2011.

[86] Cal Fussman (2006). "The Man Who Finds Planets" (http://discovermagazine.com/2006/may/cover/article_view?b_start:int=3&-C=). *Discover magazine*. . Retrieved August 13, 2007.

[87] Institute for Astronomy, University of Hawai (2010). "PS1 goes Operational and begins Science Mission, May 2010" (http://pan-starrs.ifa.hawaii.edu/public/project-status/PS1 operational.html). . Retrieved August 30, 2010.

[88] "Pan-Starrs: University of Hawaii" (http://pan-starrs.ifa.hawaii.edu/public/home.html). 2005. . Retrieved August 13, 2007.

[89] "Ice Hunters web site" (http://www.icehunters.org). Zooniverse.Org. . Retrieved July 8, 2011.

[90] "Citizen Scientists: Discover a New Horizons Flyby Target" (http://solarsystem.nasa.gov/news/display.cfm?News_ID=37726). NASA. Jun 21, 2011. . Retrieved August 23, 2011.

[91] Lakdawalla, Emily (June 21, 2011). "The most exciting citizen science project ever (to me, anyway)" (http://planetary.org/blog/article/00003073/). The Planetary Society. . Retrieved August 31, 2011.

[92] "Ice Investigators" (http://cosmoquest.org/iceinvestigators/). *web site*. CosmoQuest. 2012. . Retrieved 2012-05-23.

[93] "Finding Ice" (http://cosmoquest.org/blog/2012/05/finding-ice/). *CosmoQuest web site* (http://cosmoquest.org/). CosmoQuest. 2012-05-20. . Retrieved 2012-05-23.

[94] Kalas, Paul; Graham, James R.; Clampin, Mark C.; Fitzgerald, Michael P. (2006). "First Scattered Light Images of Debris Disks around HD 53143 and HD 139664". *The Astrophysical Journal* **637**: L57. arXiv:astro-ph/0601488. Bibcode 2006ApJ...637L..57K. doi:10.1086/500305.

[95] Trilling, D. E.; Bryden, G.; Beichman, C. A.; Rieke, G. H.; Su, K. Y. L.; Stansberry, J. A.; Blaylock, M.; Stapelfeldt, K. R.; Beeman, J. W.; Haller, E. E. (February 2008). "Debris Disks around Sun-like Stars". *The Astrophysical Journal* **674** (2): 1086–1105. arXiv:0710.5498. Bibcode 2008ApJ...674.1086T. doi:10.1086/525514.

[96] "Dusty Planetary Disks Around Two Nearby Stars Resemble Our Kuiper Belt" (http://hubblesite.org/newscenter/archive/releases/2006/05/image/a). 2006. . Retrieved July 1, 2007.

[97] Kuchner, M. J.; Stark, C. C. (2010). "Collisional Grooming Models of the Kuiper Belt Dust Cloud". *The Astronomical Journal* **140** (4): 1007–1019. arXiv:1008.0904. Bibcode 2010AJ....140.1007K. doi:10.1088/0004-6256/140/4/1007.

[98] Weissman and Johnson, 2007, *Encyclopedia of the solar system*, footnote p. 584

[99] IAU: Minor Planet Center (2011-01-03). "List Of Centaurs and Scattered-Disk Objects" (http://www.minorplanetcenter.org/iau/lists/Centaurs.html). Central Bureau for Astronomical Telegrams, Harvard-Smithsonian Center for Astrophysics. . Retrieved 2011-01-03.

External links and data sources

- Dave Jewitt's page @ UCLA (http://www2.ess.ucla.edu/~jewitt/kb.html)
  - The belt's name (http://www2.ess.ucla.edu/~jewitt/kb/gerard.html)
- Ice Hunters (http://www.icehunters.org/), a citizen science project to help discover a KBO to serve as a followup *New Horizons* mission target after Pluto
- List of short period comets by family (http://www.physics.ucf.edu/~yfernandez/cometlist.html)
- Kuiper Belt Profile (http://solarsystem.nasa.gov/planets/profile.cfm?Object=KBOs) by NASA's Solar System Exploration (http://solarsystem.nasa.gov)
- The Kuiper Belt Electronic Newsletter (http://www.boulder.swri.edu/ekonews/)
- Wm. Robert Johnston's TNO page (http://www.johnstonsarchive.net/astro/tnos.html)
- Minor Planet Center: Plot of the Outer Solar System (http://www.minorplanetcenter.org/iau/lists/OuterPlot.html), illustrating Kuiper gap
- Website of the International Astronomical Union (http://www.iau.org/) (debating the status of TNOs)
- XXVIth General Assembly 2006 (http://www.astronomy2006.com)
- nature.com article: diagram displaying inner solar system, Kuiper Belt, and Oort Cloud (http://www.nature.com/nature/journal/v424/n6949/fig_tab/nature01725_F1.html), taken from Alan Stern, S. (2003). "The evolution of comets in the Oort cloud and Kuiper belt". *Nature* **424** (6949): 639–42. doi:10.1038/nature01725. PMID 12904784.
- SPACE.com: Discovery Hints at a Quadrillion Space Rocks Beyond Neptune (http://www.space.com/scienceastronomy/060814_tno_found.html) (Sara Goudarzi) August 15, 2006 06:13 am ET
- The Outer Solar System (http://www.astronomycast.com/astronomy/episode-64-pluto-and-the-icy-outer-solar-system/) Astronomy Cast episode No. 64, includes full transcript.
- The Kuiper belt (http://365daysofastronomy.org/2009/12/08/december-8th-what-is-the-kuiper-belt/) at 365daysofastronomy.org

Trans-Neptunian object

A **trans-Neptunian object** (TNO; also written **transneptunian object**) is any object in the Solar System that orbits the Sun at a greater average distance (semi-major axis) than Neptune.

The first trans-Neptunian object to be discovered was Pluto in 1930. It took until 1992 to discover a second trans-Neptunian object orbiting the Sun directly, (15760) 1992 QB_1, with only the discovery of Pluto's moon Charon in 1978 before that. Now over 1200 trans-Neptunian objects appear on the Minor Planet Center's *List Of Transneptunian Objects*.[1] As of November 2009, two hundred of these have their orbits well-enough determined that they have been given a permanent minor planet designation.[2] [3]

The largest known trans-Neptunian objects are Eris and Pluto, followed by Makemake and Haumea. The Kuiper belt, scattered disk, and Oort cloud are three conventional divisions of this volume of space,[4] though treatments vary and a few objects such as Sedna do not fit easily into any division.[5]

History

Discovery of Pluto

The orbit of each of the planets is slightly affected by the gravitational influences of the other planets. Discrepancies in the early 1900s between the observed and expected orbits of Uranus and Neptune suggested that there were one or more additional planets beyond Neptune. The search for these led to the discovery of Pluto in 1930. However, Pluto was too small to explain the discrepancies, and revised estimates of Neptune's mass showed that the problem was spurious.

Pluto was easiest to find because it has the highest apparent magnitude of all known trans-Neptunian objects. It also has a lower inclination to the ecliptic than most other large TNOs.

Discovery of other trans-Neptunian objects

After Pluto's discovery, American astronomer Clyde Tombaugh continued searching for some years for similar objects, but found none. For a long time, no one searched for other TNOs as it was generally believed that Pluto was the only major object beyond Neptune. Only after the discovery of a second TNO, (15760) 1992 QB_1, in 1992, systematic searches for further such objects began. A broad strip of the sky around the ecliptic was photographed and digitally evaluated for slowly-moving objects. Hundreds of TNOs were found, with diameters in the range of 50 to 2500 kilometers.

Eris, at the time thought to be the largest TNO, was discovered in 2005, revisiting a long-running dispute within the scientific community over the classification of large TNOs, and whether objects like Pluto can be considered planets. Pluto and Eris were eventually classified as dwarf planets by the International Astronomical Union.

Distribution and classification

According to their distance from the Sun and their orbit parameters, TNOs are classified in two large groups:

- The (classical) **Kuiper belt**[5] contains objects with an average distance to the Sun of 30 to about 55 AU, usually having close-to-circular orbits with a small inclination from the ecliptic. Kuiper belt objects are further classified into the following two groups:
  - **Resonant objects** are locked in an orbital resonance with Neptune. Objects with a 1:2 resonance are called twotinos, and objects with a 2:3 resonance are called plutinos, after their most prominent member, Pluto.
  - **Classical Kuiper belt objects** (also called **cubewanos**) have no such resonance, moving on almost circular orbits, unperturbed by Neptune. Examples are 1992 QB_1, 50000 Quaoar and Makemake.
- The **scattered disk** contains objects farther from the Sun, usually with very irregular orbits (i.e. very elliptical and having a strong inclination from the ecliptic). A typical example is the most massive known TNO, Eris.

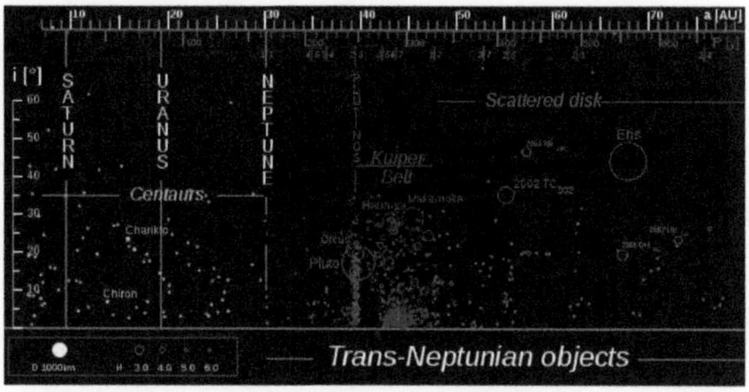

Distribution of trans-Neptunian objects

The diagram to the right illustrates the distribution of known trans-Neptunian objects (up to 70 AU) in relation to the orbits of the planets and the centaurs for reference. Different classes are represented in different colours. Resonant objects (including Neptune trojans) are plotted in red, cubewanos in blue. The scattered disk extends to the right, far beyond the diagram, with known objects at mean distances beyond 500 AU (Sedna) and aphelia beyond 1000 AU ((87269) 2000 OO_{67}).

Notable trans-Neptunian objects

- Pluto, a dwarf planet.
  - Charon, the largest of Pluto's moons.
- (15760) 1992 QB_1, the prototype cubewano, the first Kuiper belt object discovered after Pluto and Charon.
- 1998 WW_{31}, the first binary Kuiper belt object discovered after Pluto and Charon.
- (15874) 1996 TL_{66}, the first object to be identified as a scattered disc object.
- (48639) 1995 TL_8 has a very large satellite and is the earliest discovered scattered disc object.
- 1993 RO, the next plutino discovered after Pluto.
- 20000 Varuna and 50000 Quaoar, large cubewanos.
- 90482 Orcus and 28978 Ixion, large plutinos.
- 90377 Sedna, a distant object, proposed for a new category named *Extended scattered disc* (E-SDO),[6] *detached objects*,[7] *Distant Detached Objects* (DDO)[8] or *Scattered-Extended* in the formal classification by DES.[9]
- Haumea,[10] a dwarf planet, the fourth largest known trans-Neptunian object. Notable for its two known satellites and unusually short rotation period (3.9 h).[11]

10}}, Quaoar, Orcus, and Earth. These eight trans-Neptunian objects have the brightest absolute magnitudes, although several other TNOs have been found to be physically larger than Orcus, and several more may yet be found.

- Eris, a dwarf planet, a scattered disc object, and currently the most massive known trans-Neptunian object. It has one known satellite, Dysnomia.
- Makemake,[12] a dwarf planet, a cubewano, and the third largest known trans-Neptunian object.
- 2004 XR_{190}, a scattered disc object following a highly inclined but nearly circular orbit.
- (87269) 2000 OO_{67} and (148209) 2000 CR_{105}, remarkable for their eccentric orbits and large aphelia.
- 2008 KV_{42}, the first retrograde TNO, having an orbital inclination of $i = 104°$.

A fuller list of objects is being compiled in the List of trans-Neptunian objects.

Physical characteristics

Given the apparent magnitude (>20) of all but the biggest trans-Neptunian objects, the physical studies are limited to the following:

- thermal emissions for the largest objects (see size determination)
- colour indices, i.e. comparisons of the apparent magnitudes using different filters
- analysis of spectra, visual and infrared

Studying colours and spectra provides insight into the objects' origin and a potential correlation with other classes of objects, namely centaurs and some satellites of giant planets (Triton, Phoebe), suspected to originate in the Kuiper belt. However, the interpretations are typically ambiguous as the spectra can fit more than one model of the surface composition and depend on the unknown particle size. More significantly, the optical surfaces of small bodies are subject to modification by intense radiation, solar wind and micrometeorites. Consequently, the thin optical surface layer could be quite different from the regolith underneath, and not representative of the bulk composition of the body.

Small TNOs are thought to be low-density mixtures of rock and ice with some organic (carbon-containing) surface material such as tholin, detected in their spectra. On the other hand, the high density of Haumea, 2.6-3.3 g/cm^3, suggests a very high non-ice content (compare with Pluto's density: 2.0 g/cm^3).

The composition of some small TNOs could be similar to that of comets. Indeed, some centaurs undergo seasonal changes when they approach the Sun, making the boundary blurred (see 2060 Chiron and 133P/Elst–Pizarro). However, population comparisons between centaurs and TNOs are still controversial.[13]

Colours

Like centaurs, TNOs display a wide range of colours from blue-grey to very red, but unlike the centaurs, clearly re-grouped into two classes, the distribution appears to be uniform.[13]

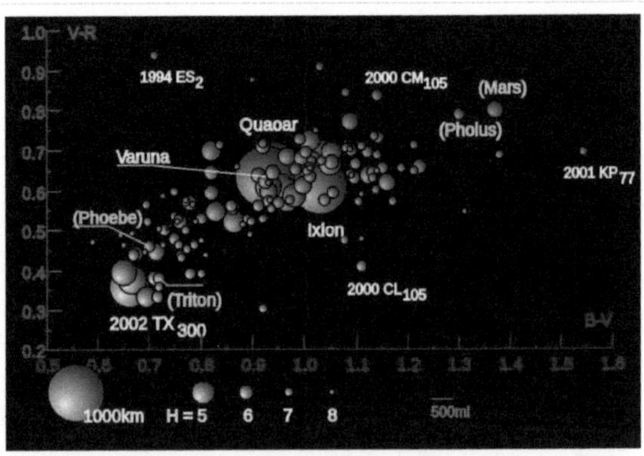

Colours of trans-Neptunian objects. Mars and Triton are not to scale. Phoebe and Pholus are not trans-Neptunian.

Colour indices are simple measures of the differences in the apparent magnitude of an object seen through blue (B), visible (V), i.e. green-yellow, and red (R) filters. The diagram illustrates known colour indices for all but the biggest objects (in slightly enhanced colour).[14] For reference, two moons: Triton and Phoebe, the centaur Pholus and the planet Mars are plotted (yellow labels, size not to scale).

Correlations between the colors and the orbital characteristics have been studied, to confirm theories of different origin of the different dynamic classes.

Classical objects

Classical objects seem to be composed of two different colour populations: the so-called cold (inclination <5°) population, displaying only red colours, and the so-called hot (higher inclination) population displaying the whole range of colours from blue to very red.[15]

A recent analysis based on the data from Deep Ecliptic Survey confirms this difference in colour between low-inclination (named *Core*) and high-inclination (named *Halo*) objects. Red colours of the Core objects together with their unperturbed orbits suggest that these objects could be a relic of the original population of the belt.[16]

Scattered disk objects

Scattered disk objects show colour resemblances with hot classical objects pointing to a common origin.

The largest objects

Characteristically, big (bright) objects are typically on inclined orbits, while the invariable plane re-groups mostly small and dim objects. While the relatively dimmer bodies, as well as the population as the whole, are reddish (V-I = 0.3–0.6), the bigger objects are often more neutral in colour (infrared index V-I < 0.2). This distinction leads to suggestion that the surface of the largest bodies is covered with ices, hiding the redder, darker areas underneath.[11]

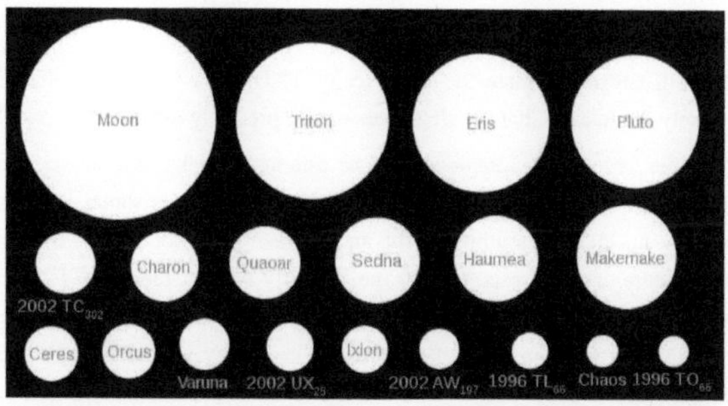

Size comparison between Earth's Moon, Neptune's moon Triton, and several large TNOs.

The diagram illustrates the relative sizes, albedos and colours of the biggest TNOs. Also shown, are the known satellites and the exceptional shape of Haumea resulting from its rapid rotation. The arc around Makemake represents uncertainty given its unknown albedo. The size of Eris follows Michael Brown's measure (2400 km) based on HST point spread model.[17] The arc around it represents the thermal measure (3000 km) by Bertoldi (see the related section of the article for the references).

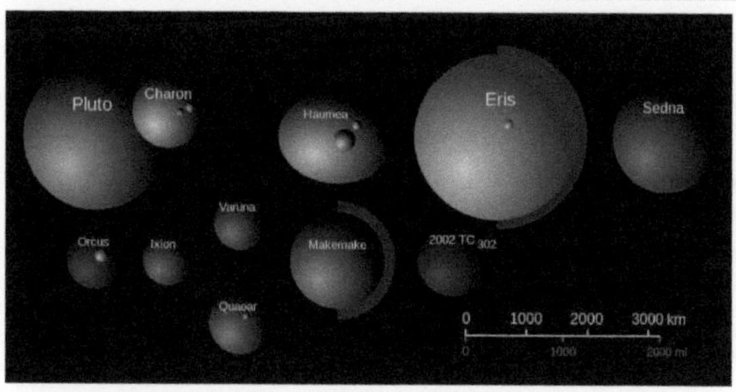

Illustration of the relative sizes, albedos and colours of the largest TNOs.

Spectra

The objects present wide range of spectra, differing in reflectivity in visible red and near infrared. Neutral objects present a flat spectrum, reflecting as much red and infrared as visible spectrum.[18] Very red objects present a steep slope, reflecting much more in red and infrared. A recent attempt at classification (common with centaurs) uses the total of four classes from **BB** (blue, average B-V=0.70, V-R=0.39 e.g. Orcus) to **RR** (very red, B-V=1.08, V-R=0.71, e.g. Sedna) with BR and IR as intermediate classes. BR and IR differ mostly in the infrared bands I, J and H.

Typical models of the surface include water ice, amorphous carbon, silicates and organic macromolecules, named tholins, created by intense radiation. Four major tholins are used to fit the reddening slope:

- Titan tholin, believed to be produced from a mixture of 90% N_2 and 10% CH_4 (gaseous methane)
- Triton tholin, as above but with very low (0.1%) methane content
- (ethane) Ice tholin I, believed to be produced from a mixture of 86% H_2O and 14% C_2H_6 (ethane)
- (methanol) Ice tholin II, 80% H_2O, 16% CH_3OH (methanol) and 3% CO_2

As an illustration of the two extreme classes **BB** and **RR**, the following compositions have been suggested

- for Sedna (**RR** very red): 24% Triton tholin, 7% carbon, 10% N_2, 26% methanol, and 33% methane
- for Orcus (**BB**, grey/blue): 85% amorphous carbon, +4% titan tholin, and 11% H_20 ice

Size determination

It is difficult to estimate the diameter of TNOs. For very large objects, with very well known orbital elements (namely, Pluto and Charon), diameters can be precisely measured by occultation of stars.

For other large TNOs, diameters can be estimated by thermal measurements. The intensity of light illuminating the object is known (from its distance to the Sun), and one assumes that most of its surface is in thermal equilibrium (usually not a bad assumption for an airless body). For a known albedo, it is possible to estimate the surface temperature, and correspondingly the intensity of heat radiation. Further, if the size of the object is known, it is possible to predict both the amount of visible light and emitted heat radiation reaching the Earth. A simplifying factor is that the Sun emits almost all of its energy in visible light and at nearby frequencies, while at the cold temperatures of TNOs, the heat radiation is emitted at completely different wavelengths (the far infrared).

Thus there are two unknowns (albedo and size), which can be determined by two independent measurements (of the amount of reflected light and emitted infrared heat radiation).

Unfortunately, TNOs are so far from the Sun that they are very cold, hence produce black-body radiation around 60 micrometres in wavelength. This wavelength of light is impossible to observe on the Earth's surface, but only from space using, e.g., the Spitzer Space Telescope. For ground-based observations, astronomers observe the tail of the black-body radiation in the far infrared. This far infrared radiation is so dim that the thermal method is only applicable to the largest KBOs. For the majority of (small) objects, the diameter is estimated by assuming an albedo. However, the albedos found range from 0.50 down to 0.05, resulting in a size range of 1200–3700 km for an object of magnitude of 1.0.[19]

See also

- Dwarf planet
- List of trans-Neptunian objects
- Mesoplanet
- Nemesis (hypothetical star)
- Small Solar System body
- Triton
- Tyche (hypothetical planet)

Notes

[1] IAU Minor Planet Center *List Of Transneptunian Objects* (http://www.minorplanetcenter.org/iau/lists/TNOs.html)

[2] List of Transneptunian objects (http://www.minorplanetcenter.org/iau/lists/TNOs.html)

[3] List of Centaurs and Scattered Disc objects (http://www.minorplanetcenter.org/iau/lists/Centaurs.html)

[4] Remo, John L. (2007). "Classifying Solid Planetary Bodies". *AIP Conference Proceedings* **886**: 284–302. Bibcode 2007AIPC..886..284R. doi:10.1063/1.2710063.

[5] The literature is inconsistent in the use of the phrases "scattered disc" and "Kuiper belt". For some, they are distinct populations; for others, the scattered disk is part of the Kuiper belt, in which case the low-eccentricity population is called the "classical Kuiper belt". Authors may even switch between these two uses in a single publication.Weissman and Johnson, 2007, *Encyclopedia of the solar system*, footnote p. 584 In this article, the scattered disk will be considered a separate population from the Kuiper belt.

[6] *Evidence for an Extended Scattered Disk?* (http://www.obs-nice.fr/gladman/cr105.html)

[7] Jewitt, D.; Delsanti, A. (2006). "The Solar System Beyond The Planets" (http://www.ifa.hawaii.edu/faculty/jewitt/papers/2006/DJ06.pdf). *Solar System Update : Topical and Timely Reviews in Solar System Sciences* (Springer-Praxis ed.). ISBN 3-540-26056-0. .

[8] Gomes, Rodney S.; Matese, John J.; Lissauer, Jack J. (2006). "A Distant Planetary-Mass Solar Companion May Have Produced Distant Detached Objects" (http://staff.on.br/rodneyg/companion/solar_companion.pdf). *Icarus* **184** (2): 589–601. Bibcode 2006Icar..184..589G. doi:10.1016/j.icarus.2006.05.026. .

[9] Elliot, J. L.; Kern, S. D.; Clancy, K. B.; Gulbis, A. A. S.; Millis, R. L.; Buie, M. W.; Wasserman, L. H.; Chiang, E. I. et al. (2005). "The Deep Ecliptic Survey: A Search for Kuiper Belt Objects and Centaurs. II. Dynamical Classification, the Kuiper Belt Plane, and the Core Population" (http://alpaca.as.arizona.edu/~trilling/des2.pdf). *The Astronomical Journal* **129** (2): 1117. Bibcode 2005AJ....129.1117E. doi:10.1086/427395. .

[10] "Distant object found orbiting Sun" (http://news.bbc.co.uk/1/hi/sci/tech/4726733.stm). *BBC News*. 2005-07-29. . Retrieved 2010-03-28.

[11] Rabinowitz, David L.; Barkume, K. M.; Brown, Michael E.; Roe, H. G.; Schwartz, M.; Tourtellotte, S. W.; Trujillo, C. A. (2006). "Photometric Observations Constraining the Size, Shape, and Albedo of 2003 El_{61}, a Rapidly Rotating, Pluto-Sized Object in the Kuiper Belt". *Astrophysical Journal* **639** (2): 1238. arXiv:astro-ph/0509401. Bibcode 2006ApJ...639.1238R. doi:10.1086/499575.

[12] http://www.minorplanetcenter.org/mpec/K05/K05O42.html

[13] Peixinho, N.; Doressoundiram, A.; Delsanti, A.; Boehnhardt, H.; Barucci, M. A.; Belskaya, I. (2003). "Reopening the TNOs Color Controversy: Centaurs Bimodality and TNOs Unimodality". *Astronomy and Astrophysics* **410** (3): L29–L32. arXiv:astro-ph/0309428. Bibcode 2003A&A...410L..29P. doi:10.1051/0004-6361:20031420.

[14] Hainaut, O. R.; Delsanti, A. C. (2002). "Color of Minor Bodies in the Outer Solar System". *Astronomy & Astrophysics* **389** (2): 641–664. Bibcode 2002A&A...389..641H. doi:10.1051/0004-6361:20020431. datasource (http://www.sc.eso.org/~ohainaut/MBOSS)

[15] Doressoundiram, A.; Peixinho, N.; de Bergh, C.; Fornasier, S.; Thébault, Ph.; Barucci, M. A.; Veillet, C.. "The color distribution in the Edgeworth-Kuiper Belt". *The Astronomical Journal* **124** (4): 2279–2296. arXiv:astro-ph/0206468. Bibcode 2002AJ....124.2279D. doi:10.1086/342447.

[16] Gulbis, Amanda A. S.; Elliot, J. L.; Kane, Julia F. (2006). "The color of the Kuiper belt Core". *Icarus* **183** (1): 168–178. Bibcode 2006Icar..183..168G. doi:10.1016/j.icarus.2006.01.021.

[17] Michael E. Brown. "The Dwarf Planets" (http://web.gps.caltech.edu/~mbrown/dwarfplanets/). California Institute of Technology, Department of Geological Sciences. . Retrieved 2008-01-26.

[18] A. Barucci *Trans Neptunian Objects' surface properties*, IAU Symposium No. 229, Asteroids, Comets, Meteors, Aug 2005, Rio de Janeiro

[19] http://www.minorplanetcenter.org/iau/lists/Sizes.html

References

External links

- Nine planets, University of Arizona (http://seds.lpl.arizona.edu/nineplanets/nineplanets/kboc.html)
- David Jewitt's Kuiper Belt site (http://www2.ess.ucla.edu/~jewitt/kb.html)
  - Large KBO page (http://www2.ess.ucla.edu/~jewitt/kb/big_kbo.html)
- A list of the estimates of the diameters from johnstonarchive (http://www.johnstonsarchive.net/astro/tnodiam.html) with references to the original papers

Sun

The Sun ☉

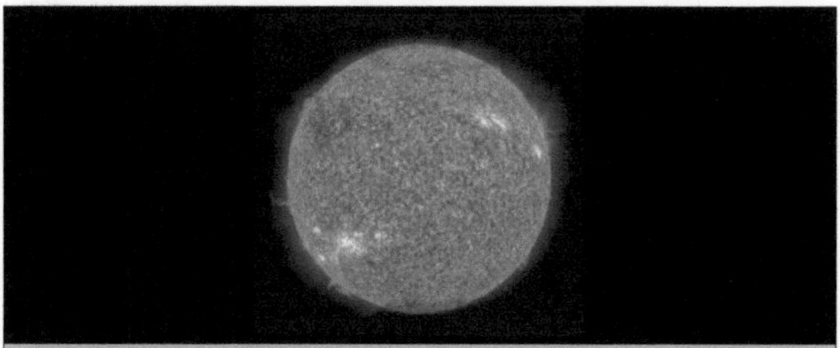

Observation data	
Mean distance from Earth	1.496×10^8 km 8 min 19 s at light speed
Visual brightness (*V*)	−26.74[1]
Absolute magnitude	4.83[1]
Spectral classification	G2V
Metallicity	$Z = 0.0122$[2]
Angular size	31.6′ – 32.7′[3]
Adjectives	Solar
Orbital characteristics	
Mean distance from Milky Way core	~2.5×10^{17} km 26000 light-years
Galactic period	$(2.25–2.50)\times10^8$ a
Velocity	~220 km/s (orbit around the center of the Galaxy) ~20 km/s (relative to average velocity of other stars in stellar neighborhood) ~370 km/s[4] (relative to the cosmic microwave background)
Physical characteristics	
Mean diameter	1.392684×10^6 km[5]
Equatorial radius	6.96342×10^5 km[5] 109 × Earth[6]
Equatorial circumference	4.379×10^6 km[6] 109 × Earth[6]
Flattening	9×10^{-6}
Surface area	6.0877×10^{12} km^2[6] 11990 × Earth[6]
Volume	1.412×10^{18} km^3[6] 1300000 × Earth

Mass	1.9891×10^{30} kg[1] $333000 \times$ Earth[1]
Average density	1.408×10^{3} kg/m^{3}[1] [6] [7]
Density	Center (model): 1.622×10^{5} kg/m^{3}[1] Lower photosphere: 2×10^{-4} kg/m^{3} Lower chromosphere: 5×10^{-6} kg/m^{3} Corona (avg): 1×10^{-12} kg/m^{3}[8]
Equatorial surface gravity	274.0 m/s^{2}[1] 27.94 *g* 27542.29 *cgs*28 × Earth[6]
Escape velocity (from the surface)	617.7 km/s[6] 55 × Earth[6]
Temperature	Center (modeled): ~1.57×10^{7} K[1] Photosphere (effective): 5778 K[1] Corona: ~5×10^{6} K
Luminosity (L_{sol})	3.846×10^{26} W[1] ~3.75×10^{28} lm ~98 lm/W efficacy
Mean intensity (I_{sol})	2.009×10^{7} W·m^{-2}·sr^{-1}
Age	4.57 billion years[9]
Rotation characteristics	
Obliquity	7.25°[1] (to the ecliptic) 67.23° (to the galactic plane)
Right ascension of North pole[10]	286.13° 19 h 4 min 30 s
Declination of North pole	+63.87° 63° 52' North
Sidereal rotation period (at equator)	25.05 days[1]
(at 16° latitude)	25.38 days[1] 25 d 9 h 7 min 12 s[10]
(at poles)	34.4 days[1]
Rotation velocity (at equator)	7.189×10^{3} km/h[6]
Photospheric composition (by mass)	
Hydrogen	73.46%[11]
Helium	24.85%
Oxygen	0.77%
Carbon	0.29%
Iron	0.16%
Neon	0.12%

Nitrogen	0.09%
Silicon	0.07%
Magnesium	0.05%
Sulfur	0.04%

The **Sun** is the star at the center of the Solar System. It is almost perfectly spherical and consists of hot plasma interwoven with magnetic fields.[12] [13] It has a diameter of about 1,392,684 km,[5] about 109 times that of Earth, and its mass (about 2×10^{30} kilograms, 330,000 times that of Earth) accounts for about 99.86% of the total mass of the Solar System.[14] Chemically, about three quarters of the Sun's mass consists of hydrogen, while the rest is mostly helium. The remainder (1.69%, which nonetheless equals 5,628 times the mass of Earth) consists of heavier elements, including oxygen, carbon, neon and iron, among others.[15]

The Sun formed about 4.6 billion years ago from the gravitational collapse of a region within a large molecular cloud. Most of the matter gathered in the center, while the rest flattened into an orbiting disk that would become the Solar System. The central mass became increasingly hot and dense, eventually initiating thermonuclear fusion in its core. It is thought that almost all other stars form by this process. The Sun's stellar classification, based on spectral class, is G2V, and is informally designated as a *yellow dwarf*, because its visible radiation is most intense in the yellow-green portion of the spectrum and although its color is white, from the surface of the Earth it may appear yellow because of atmospheric scattering of blue light.[16] In the spectral class label, *G2* indicates its surface temperature of approximately 5778 K (5505 °C), and *V* indicates that the Sun, like most stars, is a main-sequence star, and thus generates its energy by nuclear fusion of hydrogen nuclei into helium. In its core, the Sun fuses 620 million metric tons of hydrogen each second.

Once regarded by astronomers as a small and relatively insignificant star, the Sun is now thought to be brighter than about 85% of the stars in the Milky Way galaxy, most of which are red dwarfs.[17] [18] The absolute magnitude of the Sun is +4.83; however, as the star closest to Earth, the Sun is the brightest object in the sky with an apparent magnitude of −26.74.[19] [20] The Sun's hot corona continuously expands in space creating the solar wind, a stream of charged particles that extends to the heliopause at roughly 100 astronomical units. The bubble in the interstellar medium formed by the solar wind, the heliosphere, is the largest continuous structure in the Solar System.[21] [22]

The Sun is currently traveling through the Local Interstellar Cloud (near to the G-cloud) in the Local Bubble zone, within the inner rim of the Orion Arm of the Milky Way galaxy.[23] [24] Of the 50 nearest stellar systems within 17 light-years from Earth (the closest being a red dwarf named Proxima Centauri at approximately 4.2 light-years away), the Sun ranks fourth in mass.[25] The Sun orbits the center of the Milky Way at a distance of approximately 24000–26000 light-years from the galactic center, completing one clockwise orbit, as viewed from the galactic north pole, in about 225–250 million years. Since our galaxy is moving with respect to the cosmic microwave background radiation (CMB) in the direction of the constellation Hydra with a speed of 550 km/s, the Sun's resultant velocity with respect to the CMB is about 370 km/s in the direction of Crater or Leo.[26]

The mean distance of the Sun from the Earth is approximately 149.6 million kilometers (1 AU), though the distance varies as the Earth moves from perihelion in January to aphelion in July.[27] At this average distance, light travels from the Sun to Earth in about 8 minutes and 19 seconds. The energy of this sunlight supports almost all life on Earth by photosynthesis,[28] and drives Earth's climate and weather. The enormous effect of the Sun on the Earth has been recognized since prehistoric times, and the Sun has been regarded by some cultures as a deity. An accurate scientific understanding of the Sun developed slowly, and as recently as the 19th century prominent scientists had little knowledge of the Sun's physical composition and source of energy. This understanding is still developing; there are a number of present day anomalies in the Sun's behavior that remain unexplained.

Name and etymology

The English proper noun *Sun* developed from Old English *sunne* (around 725, attested in *Beowulf*), and may be related to *south*. Cognates to English *sun* appear in other Germanic languages, including Old Frisian *sunne*, *sonne*, Old Saxon *sunna*, Middle Dutch *sonne*, modern Dutch *zon*, Old High German *sunna*, modern German *Sonne*, Old Norse *sunna*, and Gothic *sunnō*. All Germanic terms for the Sun stem from Proto-Germanic **sunnōn*.[29] [30]

In relation, the Sun is personified as a goddess in Germanic paganism; Sól/Sunna.[30] Scholars theorize that the Sun, as Germanic goddess, may represent an extension of an earlier Proto-Indo-European sun deity due to Indo-European linguistic connections between Old Norse *Sól*, Sanskrit *Surya*, Gaulish *Sulis*, Lithuanian *Saulė*, and Slavic *Solntse*.[30] [31]

The English weekday name *Sunday* is attested in Old English (*Sunnandæg*; "Sun's day", from before 700) and is ultimately a result of a Germanic interpretation of Latin *dies solis*, itself a translation of the Greek *heméra helíou*.[32] The Latin name for the star, *Sol*, is widely known but is not common in general English language use; the adjectival form is the related word *solar*.[33] [34] The term *sol* is also used by planetary astronomers to refer to the duration of a solar day on another planet, such as Mars.[35] A mean Earth solar day is approximately 24 hours, while a mean Martian 'sol' is 24 hours, 39 minutes, and 35.244 seconds.[36]

Characteristics

This video takes SDO images and applies additional processing to enhance the structures visible. The events in this video represent 24 hours of activity on September 25, 2011.

The Sun is a G-type main-sequence star comprising about 99.86% of the total mass of the Solar System. It is a near-perfect sphere, with an oblateness estimated at about 9 millionths,[37] which means that its polar diameter differs from its equatorial diameter by only 10 km.[38] As the Sun consists of a plasma and is not solid, it rotates faster at its equator than at its poles. This behavior is known as differential rotation, and is caused by convection in the Sun and the movement of mass, due to steep temperature gradients from the core outwards. This mass carries a portion of the Sun's counter-clockwise angular momentum, as viewed from the ecliptic north pole, thus redistributing the angular velocity. The period of this *actual rotation* is approximately 25.6 days at the equator and 33.5 days at the poles. However, due to our constantly changing vantage point from the Earth as it orbits the Sun, the *apparent rotation* of the star at its equator is about 28 days.[39] The centrifugal effect of this slow rotation is 18 million times weaker than the surface gravity at the Sun's equator. The tidal effect of the planets is even weaker, and does not significantly affect the shape of the Sun.[40]

The Sun is a Population I, or heavy element-rich,[41] star.[42] The formation of the Sun may have been triggered by shockwaves from one or more nearby supernovae.[43] This is suggested by a high abundance of heavy elements in the Solar System, such as gold and uranium, relative to the abundances of these elements in so-called Population II (heavy element-poor) stars. These elements could most plausibly have been produced by endergonic nuclear reactions during a supernova, or by transmutation through neutron absorption inside a massive second-generation star.[42]

The Sun does not have a definite boundary as rocky planets do, and in its outer parts the density of its gases drops exponentially with increasing distance from its center.[44] Nevertheless, it has a well-defined interior structure, described below. The Sun's radius is measured from its center to the edge of the photosphere. This is simply the layer above which the gases are too cool or too thin to radiate a significant amount of light, and is therefore the surface most readily visible to the naked eye.[39]

The solar interior is not directly observable, and the Sun itself is opaque to electromagnetic radiation. However, just as seismology uses waves generated by earthquakes to reveal the interior structure of the Earth, the discipline of helioseismology makes use of pressure waves (infrasound) traversing the Sun's interior to measure and visualize the star's inner structure.[39] Computer modeling of the Sun is also used as a theoretical tool to investigate its deeper layers.

Core

The core of the Sun is considered to extend from the center to about 20–25% of the solar radius.[45] It has a density of up to 150 g/cm^3[46] [47] (about 150 times the density of water) and a temperature of close to 15.7 million kelvin (K). By contrast, the Sun's surface temperature is approximately 5,800 K. Recent analysis of SOHO mission data favors a faster rotation rate in the core than in the rest of the radiative zone.[45] Through most of the Sun's life, energy is produced by nuclear fusion through a series of steps called the p–p (proton–proton) chain; this process converts hydrogen into helium.[48] Only 0.8% of the energy generated in the Sun comes from the CNO cycle.[49]

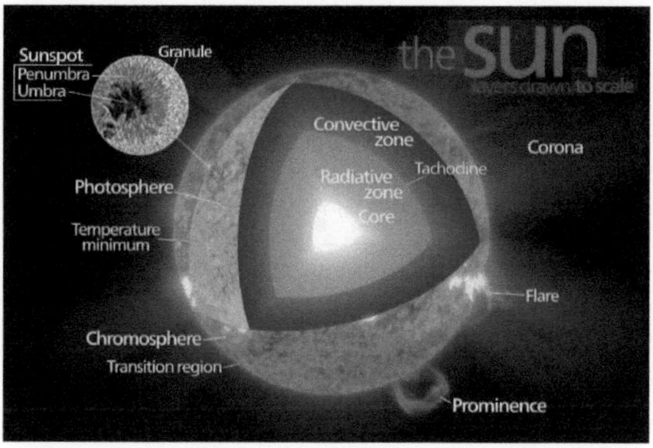

The structure of the sun

The core is the only region in the Sun that produces an appreciable amount of thermal energy through fusion; inside 24% of the Sun's radius, 99% of the power has been generated, and by 30% of the radius, fusion has stopped nearly entirely. The rest of the star is heated by energy that is transferred outward from the core and the layers just outside. The energy produced by fusion in the core must then travel through many successive layers to the solar photosphere before it escapes into space as sunlight or kinetic energy of particles.[44] [39]

The proton–proton chain occurs around 9.2×10^{37} times each second in the core of the Sun. Since this reaction uses four free protons (hydrogen nuclei), it converts about 3.7×10^{38} protons to alpha particles (helium nuclei) every second (out of a total of ~8.9×10^{56} free protons in the Sun), or about 6.2×10^{11} kg per second.[39] Since fusing hydrogen into helium releases around 0.7% of the fused mass as energy,[50] the Sun releases energy at the mass-energy conversion rate of 4.26 million metric tons per second, 384.6 yotta watts (3.846×10^{26} W),[1] or 9.192×10^{10} megatons of TNT per second. This mass is not destroyed to create the energy, rather, the mass is carried away *in* the radiated energy, as described by the concept of mass-energy equivalence.

The power production by fusion in the core varies with distance from the solar center. At the center of the Sun, theoretical models estimate it to be approximately 276.5 watts/m^3,[51] a power production density that more nearly approximates reptile metabolism than a thermonuclear bomb.[52] Peak power production in the Sun has been compared to the volumetric heats generated in an active compost heap. The tremendous power output of the Sun is not due to its high power per volume, but instead due to its large size.

The fusion rate in the core is in a self-correcting equilibrium: a slightly higher rate of fusion would cause the core to heat up more and expand slightly against the weight of the outer layers, reducing the fusion rate and correcting the perturbation; and a slightly lower rate would cause the core to cool and shrink slightly, increasing the fusion rate and again reverting it to its present level.[53] [54]

The gamma rays (high-energy photons) released in fusion reactions are absorbed in only a few millimeters of solar plasma and then re-emitted again in random direction and at slightly lower energy. Therefore it takes a long time for radiation to reach the Sun's surface. Estimates of the photon travel time range between 10,000 and 170,000 years.[55] In contrast, it takes only 2.3 seconds for the neutrinos, which account for about 2% of the total energy production of the Sun, to reach the surface. Since energy transport in the Sun is a process which involves photons in thermodynamic equilibrium with matter, the time scale of energy transport in the Sun is longer, on the order of 30,000,000 years. This is the time it would take the Sun to return to a stable state if the rate of energy generation in its core were suddenly to be changed.[56]

After a final trip through the convective outer layer to the transparent surface of the photosphere, the photons escape as visible light. Each gamma ray in the Sun's core is converted into several million photons of visible light before escaping into space. Neutrinos are also released by the fusion reactions in the core, but unlike photons they rarely interact with matter, so almost all are able to escape the Sun immediately. For many years measurements of the number of neutrinos produced in the Sun were lower than theories predicted by a factor of 3. This discrepancy was resolved in 2001 through the discovery of the effects of neutrino oscillation: the Sun emits the number of neutrinos predicted by the theory, but neutrino detectors were missing $^2/_3$ of them because the neutrinos had changed flavor by the time they were detected.[57]

Radiative zone

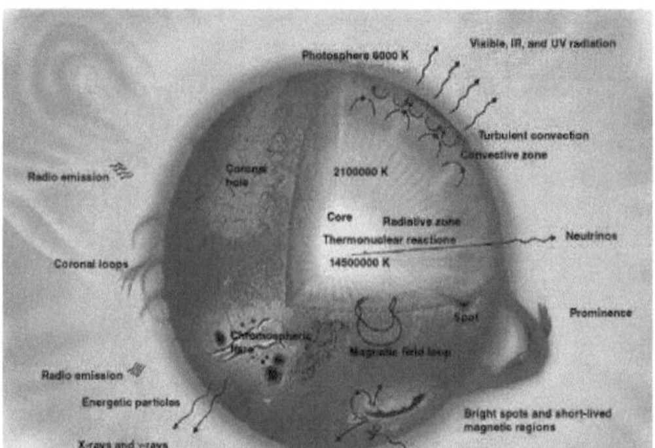

Cross-section of a solar-type star (NASA)

Below about 0.7 solar radii, solar material is hot and dense enough that thermal radiation is sufficient to transfer the intense heat of the core outward.[58] This zone is not regulated by thermal convection; however the temperature drops from approximately 7 to 2 million kelvin with increasing altitude.[47] This temperature gradient is less than the value of the adiabatic lapse rate and hence cannot drive convection.[47] Energy is transferred by radiation—ions of hydrogen and helium emit photons, which travel only a brief distance before being reabsorbed by other ions.[58] The density drops a hundredfold (from 20 g/cm^3 to only 0.2 g/cm^3) from 0.25 solar radii to the top of the radiative zone.[58]

The radiative zone and the convection form a transition layer, the tachocline. This is a region where the sharp regime change between the uniform rotation of the radiative zone and the differential rotation of the convection zone results in a large shear—a condition where successive horizontal layers slide past one another.[59] The fluid motions found in the convection zone above, slowly disappear from the top of this layer to its bottom, matching the calm characteristics of the radiative zone on the bottom. Presently, it is hypothesized (see Solar dynamo), that a magnetic dynamo within this layer generates the Sun's magnetic field.[47]

Convective zone

In the Sun's outer layer, from its surface to approximately 200,000 km below (70% of the solar radius away from the center), the solar plasma is not dense enough or hot enough to transfer the thermal energy of the interior outward through radiation; in other words it is opaque enough. As a result, thermal convection occurs as thermal columns carry hot material to the surface (photosphere) of the Sun. Once the material cools off at the surface, it plunges downward to the base of the convection zone, to receive more heat from the top of the radiative zone. At the visible surface of the Sun, the temperature has dropped to 5,700 K and the density to only 0.2 g/m^3 (about 1/6,000th the density of air at sea level).[47]

The thermal columns in the convection zone form an imprint on the surface of the Sun as the solar granulation and supergranulation. The turbulent convection of this outer part of the solar interior causes a "small-scale" dynamo that produces magnetic north and south poles all over the surface of the Sun.[47] The Sun's thermal columns are Bénard cells and therefore tend to be hexagonal prisms.[60]

Photosphere

The visible surface of the Sun, the photosphere, is the layer below which the Sun becomes opaque to visible light.[61] Above the photosphere visible sunlight is free to propagate into space, and its energy escapes the Sun entirely. The change in opacity is due to the decreasing amount of H^- ions, which absorb visible light easily.[61] Conversely, the visible light we see is produced as electrons react with hydrogen atoms to produce H^- ions.[62] [63] The photosphere is tens to hundreds of kilometers thick, being slightly less opaque than air on Earth. Because the upper part of the photosphere is cooler than the lower part, an image of the Sun appears brighter in the center than on the edge or *limb* of the solar disk, in a phenomenon known as limb darkening.[61] Sunlight has approximately a black-body spectrum that indicates its temperature is about 6,000 K, interspersed with atomic absorption lines from the tenuous layers above the photosphere. The photosphere has a particle density of ~10^{23} m^{-3}. (This is about 0.37% of the particle number per volume of Earth's atmosphere at sea level.) The photosphere is not fully ionized—the extent of ionization is about 3%, leaving almost all of the hydrogen in atomic form.[64]

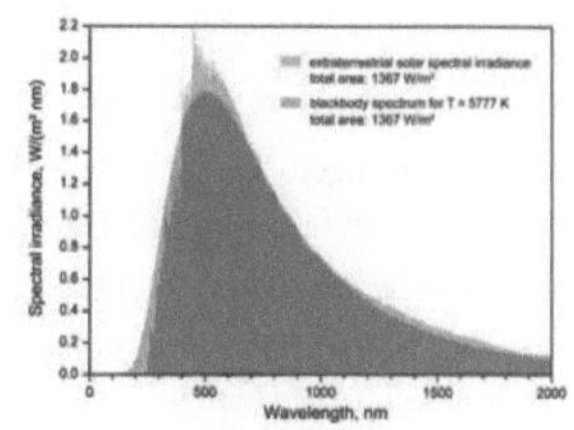

The effective temperature, or black body temperature, of the Sun (5777 K) is the temperature a black body of the same size must have to yield the same total emissive power.

During early studies of the optical spectrum of the photosphere, some absorption lines were found that did not correspond to any chemical elements then known on Earth. In 1868, Norman Lockyer hypothesized that these absorption lines were because of a new element which he dubbed *helium*, after the Greek Sun god Helios. It was not until 25 years later that helium was isolated on Earth.[65]

Atmosphere

During a total solar eclipse, the solar corona can be seen with the naked eye, during the brief period of totality.

The parts of the Sun above the photosphere are referred to collectively as the *solar atmosphere*.[61] They can be viewed with telescopes operating across the electromagnetic spectrum, from radio through visible light to gamma rays, and comprise five principal zones: the *temperature minimum*, the chromosphere, the transition region, the corona, and the heliosphere.[61] The heliosphere, which may be considered the tenuous outer atmosphere of the Sun, extends outward past the orbit of Pluto to the heliopause, where it forms a sharp shock front boundary with the interstellar medium. The chromosphere, transition region, and corona are much hotter than the surface of the Sun.[61] The reason has not been conclusively proven; evidence suggests that Alfvén waves may have enough energy to heat the corona.[66]

The coolest layer of the Sun is a temperature minimum region about 500 km above the photosphere, with a temperature of about 4100 K.[61] This part of the Sun is cool enough to support simple molecules such as carbon monoxide and water, which can be detected by their absorption spectra.[67]

Above the temperature minimum layer is a layer about 2000 km thick, dominated by a spectrum of emission and absorption lines.[61] It is called the *chromosphere* from the Greek root *chroma*, meaning color, because the chromosphere is visible as a colored flash at the beginning and end of total eclipses of the Sun.[58] The temperature in the chromosphere increases gradually with altitude, ranging up to around 20000 K near the top.[61] In the upper part of chromosphere helium becomes partially ionized.[68]

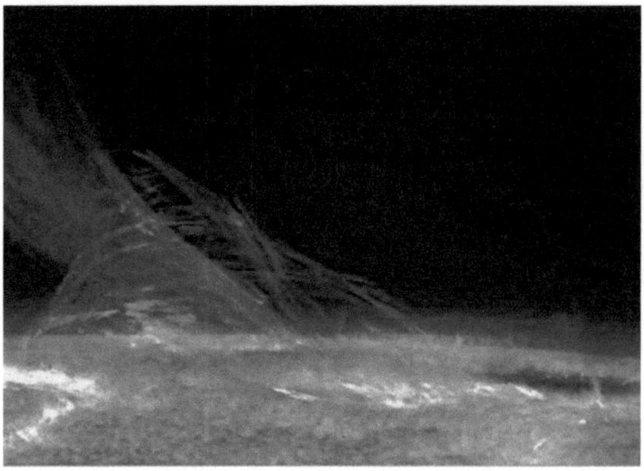

Taken by Hinode's Solar Optical Telescope on January 12, 2007, this image of the Sun reveals the filamentary nature of the plasma connecting regions of different magnetic polarity.

Above the chromosphere, in a thin (about 200 km) transition region, the temperature rises rapidly from around 20,000 K in the upper chromosphere to coronal temperatures closer to 1,000,000 K.[69] The temperature increase is facilitated by the full ionization of helium in the transition region, which significantly reduces radiative cooling of the plasma.[68] The transition region does not occur at a well-defined altitude. Rather, it forms a kind of nimbus around chromospheric features such as spicules and filaments, and is in constant, chaotic motion.[58] The transition region is not easily visible from Earth's surface, but is readily observable from space by instruments sensitive to the extreme ultraviolet portion of the spectrum.[70]

The corona is the extended outer atmosphere of the Sun, which is much larger in volume than the Sun itself. The corona continuously expands into space forming the solar wind, which fills all the Solar System.[71] The low corona, near the surface of the Sun, has a particle density around 10^{15}–10^{16} m^{-3}.[68] [72] The average temperature of the corona and solar wind is about 1,000,000–2,000,000 K; however, in the hottest regions it is 8,000,000–20,000,000

K.[69] While no complete theory yet exists to account for the temperature of the corona, at least some of its heat is known to be from magnetic reconnection.[69] [71]

The heliosphere, which is the cavity around the Sun filled with the solar wind plasma, extends from approximately 20 solar radii (0.1 AU) to the outer fringes of the Solar System. Its inner boundary is defined as the layer in which the flow of the solar wind becomes *superalfvénic*—that is, where the flow becomes faster than the speed of Alfvén waves.[73] Turbulence and dynamic forces outside this boundary cannot affect the shape of the solar corona within, because the information can only travel at the speed of Alfvén waves. The solar wind travels outward continuously through the heliosphere, forming the solar magnetic field into a spiral shape,[71] until it impacts the heliopause more than 50 AU from the Sun. In December 2004, the Voyager 1 probe passed through a shock front that is thought to be part of the heliopause. Both of the Voyager probes have recorded higher levels of energetic particles as they approach the boundary.[74]

Magnetic field

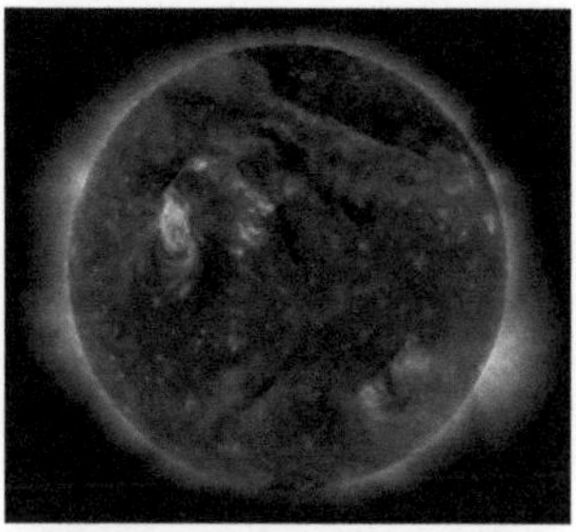

In this false-color ultraviolet image, the Sun shows a C3-class solar flare (white area on upper left), a solar tsunami (wave-like structure, upper right) and multiple filaments of plasma following a magnetic field, rising from the stellar surface.

The Sun is a magnetically active star. It supports a strong, changing magnetic field that varies year-to-year and reverses direction about every eleven years around solar maximum.[44] The Sun's magnetic field leads to many effects that are collectively called solar activity, including sunspots on the surface of the Sun, solar flares, and variations in solar wind that carry material through the Solar System.[44] Effects of solar activity on Earth include auroras at moderate to high latitudes, and the disruption of radio communications and electric power. Solar activity is thought to have played a large role in the formation and evolution of the Solar System. Solar activity changes the structure of Earth's outer atmosphere.[39]

All matter in the Sun is in the form of gas and plasma because of its high temperatures. This makes it possible for the Sun to rotate faster at its equator (about 25 days) than it does at higher latitudes (about 35 days near its poles). The differential rotation of the Sun's latitudes causes its magnetic field lines to become twisted together over time, causing magnetic field loops to erupt from the Sun's surface and trigger the formation of the Sun's dramatic sunspots and solar prominences (see magnetic reconnection). This twisting action creates the solar dynamo and an 11-year solar cycle of magnetic activity as the Sun's magnetic field reverses itself about every 11 years.[76] [77]

The heliospheric current sheet extends to the outer reaches of the Solar System, and results from the influence of the Sun's rotating magnetic field on the plasma in the interplanetary medium.[75]

The solar magnetic field extends well beyond the Sun itself. The magnetized solar wind plasma carries Sun's magnetic field into the space forming what is called the interplanetary magnetic field.[71] Since the plasma can only move along the magnetic field lines, the interplanetary magnetic field is initially stretched radially away from the Sun. Because the fields above and below the solar equator have different polarities pointing towards and away from the Sun, there exists a thin current layer in the solar equatorial plane, which is called the heliospheric current sheet.[71] At the large distances the rotation of the Sun twists the magnetic field and the current sheet into the Archimedean spiral like structure called the Parker spiral.[71] The interplanetary magnetic field is much stronger than the dipole component of the solar magnetic field. The Sun's dipole magnetic field of 50–400 μT (at the photosphere) reduces with the cube of the distance to about 0.1 nT at the distance of the Earth. However, according to spacecraft observations the interplanetary field at the Earth's location is around 5 nT, about a hundred times greater.[78] The difference is due to magnetic fields generated by electrical currents in the plasma surrounding the sun.

Chemical composition

The Sun is composed primarily of the chemical elements hydrogen and helium; they account for 74.9% and 23.8% of the mass of the Sun in the photosphere, respectively.[79] All heavier elements, called *metals* in astronomy, account for less than 2% of the mass. The most abundant metals are oxygen (roughly 1% of the Sun's mass), carbon (0.3%), neon (0.2%), and iron (0.2%).[80]

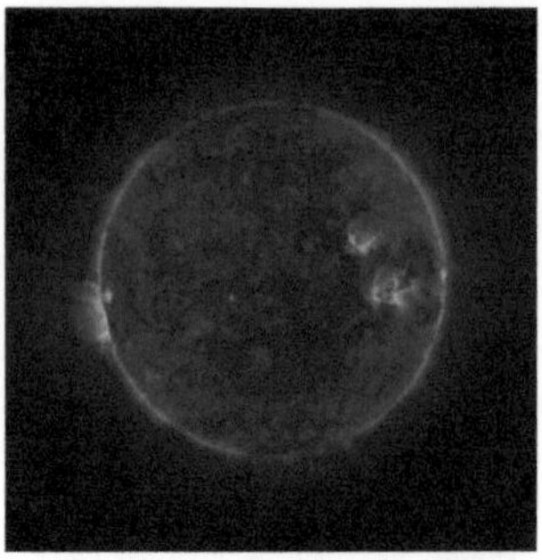

Stereo Image taken by NASA STEREO Probes launched in 2006; utilizing two Spacecrafts to image the Sun at the extreme UV wavelength (171 Å).

The Sun inherited its chemical composition from the interstellar medium out of which it formed: the hydrogen and helium in the Sun were produced by Big Bang nucleosynthesis. The metals were produced by stellar nucleosynthesis in generations of stars which completed their stellar evolution and returned their material to the interstellar medium before the formation of the Sun.[81] The chemical composition of the photosphere is normally considered representative of the composition of the primordial Solar System.[82] However, since the Sun formed, the helium and heavy elements have settled out of the photosphere. Therefore, the photosphere now contains slightly less helium and only 84% of the heavy elements than the protostellar Sun did; the protostellar Sun was 71.1% hydrogen, 27.4% helium, and 1.5% metals.[79]

In the inner portions of the Sun, nuclear fusion has modified the composition by converting hydrogen into helium, so the innermost portion of the Sun is now roughly 60% helium, with the metal abundance unchanged. Because the interior of the Sun is radiative, not convective (see Radiative zone above), none of the fusion products from the core have risen to the photosphere.[83]

The solar heavy-element abundances described above are typically measured both using spectroscopy of the Sun's photosphere and by measuring abundances in meteorites that have never been heated to melting temperatures. These meteorites are thought to retain the composition of the protostellar Sun and thus not affected by settling of heavy elements. The two methods generally agree well.[15]

Singly ionized iron group elements

In the 1970s, much research focused on the abundances of iron group elements in the Sun.[84] [85] Although significant research was done, the abundance determination of some iron group elements (e.g., cobalt and manganese) was still difficult at least as far as 1978 because of their hyperfine structures.[84]

The first largely complete set of oscillator strengths of singly ionized iron group elements were made available first in the 1960s,[86] and improved oscillator strengths were computed in 1976.[87] In 1978 the abundances of 'Singly Ionized' elements of the iron group were derived.[84]

Solar and planetary mass fractionation relationship

Various authors have considered the existence of a mass fractionation relationship between the isotopic compositions of solar and planetary noble gases,[88] for example correlations between isotopic compositions of planetary and solar neon and xenon.[89] Nevertheless, the belief that the whole Sun has the same composition as the solar atmosphere was still widespread, at least until 1983.[90]

In 1983, it was claimed that it was the fractionation in the Sun itself that caused the fractionation relationship between the isotopic compositions of planetary and solar wind implanted noble gases.[90]

Solar cycles

Sunspots and the sunspot cycle

When observing the Sun with appropriate filtration, the most immediately visible features are usually its sunspots, which are well-defined surface areas that appear darker than their surroundings because of lower temperatures. Sunspots are regions of intense magnetic activity where convection is inhibited by strong magnetic fields, reducing energy transport from the hot interior to the surface. The magnetic field causes strong heating in the corona, forming active regions that are the source of intense solar flares and coronal mass ejections. The largest sunspots can be tens of thousands of kilometers across.[91]

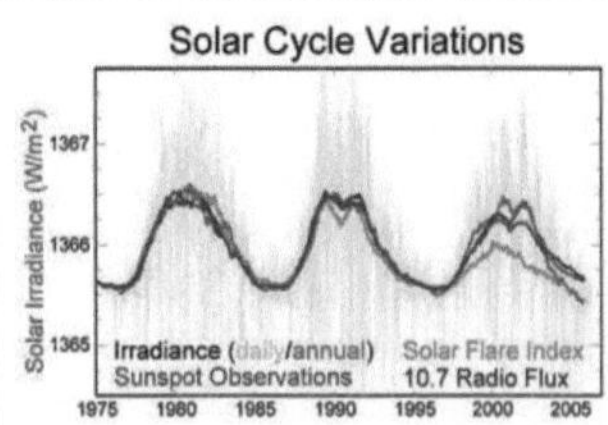

Measurements of solar cycle variation during the last 30 years

The number of sunspots visible on the Sun is not constant, but varies over an 11-year cycle known as the solar cycle. At a typical solar minimum, few sunspots are visible, and occasionally none at all can be seen. Those that do appear are at high solar latitudes. As the sunspot cycle progresses, the number of sunspots increases and they move closer to the equator of the Sun, a phenomenon described by Spörer's law. Sunspots usually exist as pairs with opposite magnetic polarity. The magnetic polarity of the leading sunspot alternates every solar cycle, so that it will be a north magnetic pole in one solar cycle and a south magnetic pole in the next.[92]

The solar cycle has a great influence on space weather, and a significant influence on the Earth's climate since the Sun's luminosity has a direct relationship with magnetic activity.[93] Solar activity minima tend to be correlated with colder temperatures, and longer than average solar cycles tend to be correlated with hotter temperatures. In the 17th century, the solar cycle appeared to have stopped entirely for several decades; few sunspots were observed during this period. During this era, known as the Maunder minimum or Little Ice Age, Europe experienced unusually cold temperatures.[94] Earlier extended minima have been discovered through analysis of tree rings and appear to have coincided with lower-than-average global temperatures.[95]

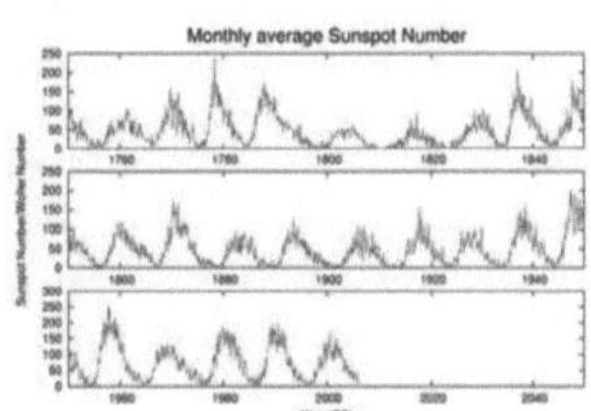

History of the number of observed sunspots during the last 250 years, which shows the ~11-year solar cycle

Possible long-term cycle

A recent theory claims that there are magnetic instabilities in the core of the Sun that cause fluctuations with periods of either 41,000 or 100,000 years. These could provide a better explanation of the ice ages than the Milankovitch cycles.[96] [97]

Life phases

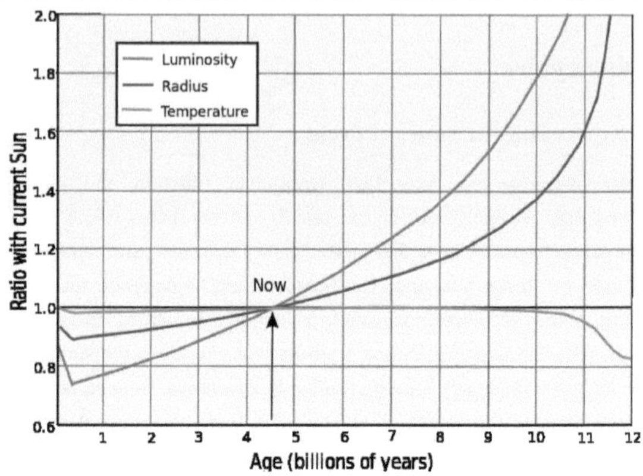

Evolution of the Sun's luminosity, radius and effective temperature compared to the present Sun. After Ribas (2010)[98]

The Sun was formed about 4.57 billion years ago from the collapse of part of a giant molecular cloud that consisted mostly of hydrogen and helium and which probably gave birth to many other stars.[44] This age is estimated using computer models of stellar evolution and through nucleocosmochronology.[9] The result is consistent with the radiometric date of the oldest Solar System material, at 4.567 billion years ago.[99] [100] Studies of ancient meteorites reveal traces of stable daughter nuclei of short-lived isotopes, such as iron-60, that form only in exploding, short-lived stars. This indicates that one or more supernovae must have occurred near the location where the Sun formed. A shock wave from a nearby supernova would have triggered the formation of the Sun by compressing the gases within the molecular cloud, and causing certain regions to collapse under their own gravity.[101] As one fragment of the cloud collapsed it also began to rotate due to conservation of angular momentum and heat up with the increasing pressure. Much of the mass became concentrated in the center, while the rest flattened out into a disk which would become the planets and other solar system bodies. Gravity and pressure within the core of the cloud generated a lot of heat as it accreted more gas from the surrounding disk, eventually triggering nuclear fusion. Thus, the Sun was born.

The Sun is about halfway through its main-sequence stage, during which nuclear fusion reactions in its core fuse hydrogen into helium. Each second, more than four million tonnes of matter are converted into energy within the Sun's core, producing neutrinos and solar radiation. At this rate, the Sun has so far converted around 100 Earth-masses of matter into energy. The Sun will spend a total of approximately 10 billion years as a main-sequence star.[102]

The Sun does not have enough mass to explode as a supernova. Instead, in about 5 billion years, it will enter a red giant phase. Its outer layers will expand as the hydrogen fuel at the core is consumed and the core will contract and heat up. Hydrogen fusion will continue along a shell surrounding a helium core, which will steadily expand as more helium is produced. Once the core temperature reaches around 100 million kelvin, helium fusion at the core will begin producing carbon, and the Sun will enter the asymptotic giant branch phase.[42] Following the red giant phase, intense thermal pulsations will cause the Sun to throw off its outer layers, forming a planetary nebula. The only object that will remain after the outer layers are ejected is the extremely hot stellar core, which will slowly cool and fade as a white dwarf over many billions of years. This stellar evolution scenario is typical of low- to medium-mass stars.[103] [104]

Earth's fate

As a red giant, the Sun will have a maximum radius beyond the Earth's current orbit, 1 AU (1.5×10^{11} m), 250 times the present radius of the Sun.[105] When the Sun is an asymptotic giant branch star, it will have lost roughly 30% of its present mass due to a stellar wind, so the orbits of the planets will move outward. If it were only for this, Earth would probably be unaffected. However, new research suggests that Earth will be swallowed by the Sun owing to tidal interactions.[105] If Earth should escape incineration in the Sun, its water will be boiled away and most of its atmosphere will escape into space. During its life in the main sequence, the Sun is becoming more luminous (about 10% every 1 billion years) and its surface temperature is slowly rising. The Sun used to be fainter in its early past; which is possibly the reason life on Earth has only existed for two to three billion years. The increase in solar temperatures is such that in about another billion years the surface of the Earth will probably become too hot for liquid water to exist, ending all terrestrial life.[105] [106]

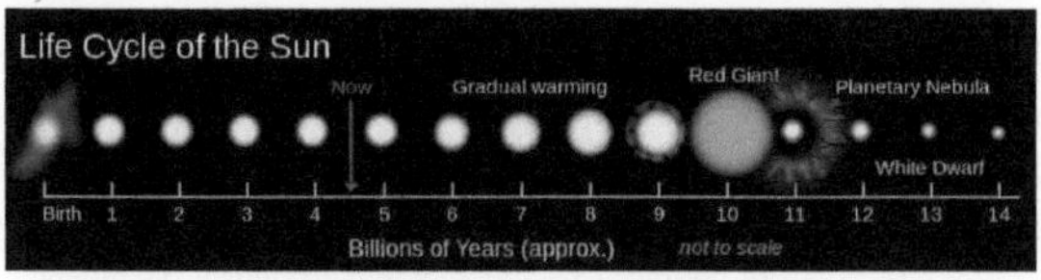

Life-cycle of the Sun; sizes are not drawn to scale.

Sunlight

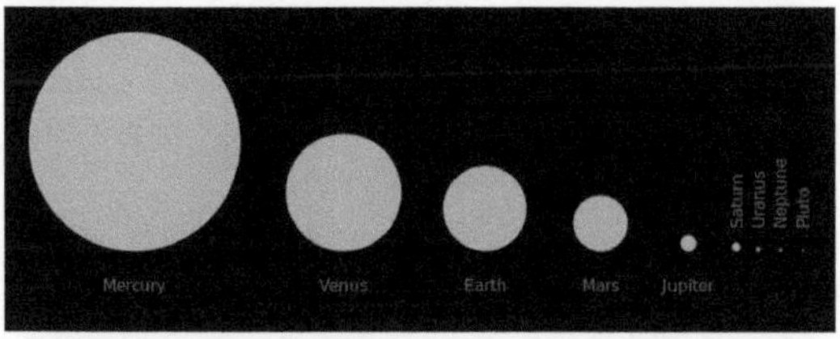

Comparison of the Sun's apparent size, as seen from the vicinity of Mercury, Venus, Earth, Mars, Jupiter, Saturn, Uranus, Neptune and Pluto

Sunlight is Earth's primary source of energy. The only other source of energy the earth has are the fissionable materials generated by the cataclysmic death of another star. These fissionable materials trapped in the Earth's crust is what gives rise to geothermal energy, which drives the volcanism on Earth while also making it possible for mankind to fuel nuclear reactors. The solar constant is the amount of power that the Sun deposits per unit area that is directly exposed to sunlight. The solar constant is equal to approximately 1368 W/m^2 (watts per square meter) at a distance of one astronomical unit (AU) from the Sun (that is, on or near Earth).[107] Sunlight at the top of Earth's atmosphere is composed (by total energy) of about 50% infrared light, 40% visible light, and 10% ultraviolet light.[108]

Sunlight on the surface of Earth is attenuated by the Earth's atmosphere so that less power arrives at the surface—closer to 1000 W/m^2 in clear conditions when the Sun is near the zenith.[109] The atmosphere in particular filters out over 70% of solar ultraviolet, especially at the shorter wavelengths.[110]

Solar energy can be harnessed by a variety of natural and synthetic processes—photosynthesis by plants captures the energy of sunlight and converts it to chemical form (oxygen and reduced carbon compounds), while direct heating or electrical conversion by solar cells are used by solar power equipment to generate electricity or to do other useful work, sometimes employing concentrating solar power (that it is measured in suns). The energy stored in petroleum and other fossil fuels was originally converted from sunlight by photosynthesis in the distant past.[39]

The Sun as it appears from the surface of the Earth at Sunset.

Motion and location within the galaxy

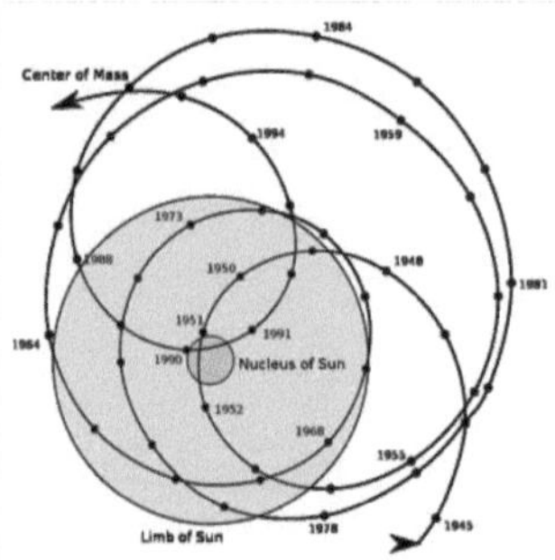

Motion of the barycenter of the Solar System relative to the Sun

The Sun lies close to the inner rim of the Milky Way Galaxy's Orion Arm, in the Local Fluff or the Gould Belt, at a hypothesized distance of 7.5–8.5 kpc (25,000–28,000 lightyears) from the Galactic Center,[111] [112] [113] [114] contained within the Local Bubble, a space of rarefied hot gas, possibly produced by the supernova remnant, Geminga.[115] The distance between the local arm and the next arm out, the Perseus Arm, is about 6,500 light-years.[116] The Sun, and thus the Solar System, is found in what scientists call the galactic habitable zone.

The Apex of the Sun's Way, or the solar apex, is the direction that the Sun travels through space in the Milky Way, relative to other nearby stars. The general direction of the Sun's galactic motion is towards the star Vega in the constellation of Lyra at an angle of roughly 60 sky degrees to the direction of the Galactic Center.

Illustration of the Milky Way galaxy, showing the location of the Sun

The Sun's orbit around the Galaxy is expected to be roughly elliptical with the addition of perturbations due to the galactic spiral arms and non-uniform mass distributions. In addition the Sun oscillates up and down relative to the galactic plane approximately 2.7 times per orbit. It has been argued that the Sun's passage through the higher density spiral arms often coincides with mass extinctions on Earth, perhaps due to increased impact events.[117] It takes the Solar System about 225–250 million years to complete one orbit of the galaxy (a *galactic year*),[118] so it is thought to have completed 20–25 orbits during the lifetime of the Sun. The orbital speed of the Solar System about the center of the Galaxy is approximately 251 km/s.[119] At this speed, it takes around 1,190 years for the Solar System to travel a distance of 1 light-year, or 7 days to travel 1 AU.[120]

The Sun's motion about the center of mass of the Solar System is complicated by perturbations from the planets. Every few hundred years this motion switches between prograde and retrograde.[121]

Theoretical problems

Solar neutrino problem

For many years the number of solar electron neutrinos detected on Earth was ⅓ to ½ of the number predicted by the standard solar model. This anomalous result was termed the solar neutrino problem. Theories proposed to resolve the problem either tried to reduce the temperature of the Sun's interior to explain the lower neutrino flux, or posited that electron neutrinos could oscillate—that is, change into undetectable tau and muon neutrinos as they traveled between the Sun and the Earth.[122] Several neutrino observatories were built in the 1980s to measure the solar neutrino flux as accurately as possible, including the Sudbury Neutrino Observatory in Canada and the Kamiokande laboratory in Japan.[123] Results from these observatories eventually led to the discovery that neutrinos have a very small rest mass and do indeed oscillate.[124] [57] Moreover, in 2001 the Sudbury Neutrino Observatory was able to detect all three types of neutrinos directly, and found that the Sun's *total* neutrino emission rate agreed with the Standard Solar Model, although depending on the neutrino energy as few as one-third of the neutrinos seen at Earth are of the electron type.[123] [125] This proportion agrees with that predicted by the Mikheyev–Smirnov–Wolfenstein effect (also known as the matter effect), which describes neutrino oscillation in matter, and it is now considered a solved problem.[123]

Coronal heating problem

The optical surface of the Sun (the photosphere) is known to have a temperature of approximately 6,000 K. Above it lies the solar corona, rising to a temperature of 1,000,000–2,000,000 K.[69] The high temperature of the corona shows that it is heated by something other than direct heat conduction from the photosphere.[71]

It is thought that the energy necessary to heat the corona is provided by turbulent motion in the convection zone below the photosphere, and two main mechanisms have been proposed to explain coronal heating.[69] The first is wave heating, in which sound, gravitational or magnetohydrodynamic waves are produced by turbulence in the convection zone.[69] These waves travel upward and dissipate in the corona, depositing their energy in the ambient gas in the form of heat.[126] The other is magnetic heating, in which magnetic energy is continuously built up by photospheric motion and released through magnetic reconnection in the form of large solar flares and myriad similar but smaller events—nanoflares.[127]

Currently, it is unclear whether waves are an efficient heating mechanism. All waves except Alfvén waves have been found to dissipate or refract before reaching the corona.[128] In addition, Alfvén waves do not easily dissipate in the corona. Current research focus has therefore shifted towards flare heating mechanisms.[69]

Faint young Sun problem

Theoretical models of the Sun's development suggest that 3.8 to 2.5 billion years ago, during the Archean period, the Sun was only about 75% as bright as it is today. Such a weak star would not have been able to sustain liquid water on the Earth's surface, and thus life should not have been able to develop. However, the geological record demonstrates that the Earth has remained at a fairly constant temperature throughout its history, and that the young Earth was somewhat warmer than it is today. The consensus among scientists is that the young Earth's atmosphere contained much larger quantities of greenhouse gases (such as carbon dioxide, methane and/or ammonia) than are present today, which trapped enough heat to compensate for the smaller amount of solar energy reaching the planet.[129]

History of observation

Early understanding

The Trundholm Sun chariot pulled by a horse is a sculpture believed to be illustrating an important part of Nordic Bronze Age mythology. The sculpture is probably from around 1350 BC. It is displayed at the National Museum of Denmark.

Like other natural phenomena, the Sun has been an object of veneration in many cultures throughout human history. Humanity's most fundamental understanding of the Sun is as the luminous disk in the sky, whose presence above the horizon creates day and whose absence causes night. In many prehistoric and ancient cultures, the Sun was thought to be a solar deity or other supernatural phenomenon. Worship of the Sun was central to civilizations such as the Inca of South America and the Aztecs of what is now Mexico. Many ancient monuments were constructed with solar phenomena in mind; for example, stone megaliths accurately mark the summer or winter solstice (some of the most prominent megaliths are located in Nabta Playa, Egypt; Mnajdra, Malta and at Stonehenge, England); Newgrange, a prehistoric human-built mount in Ireland, was designed to detect the winter solstice; the pyramid of El Castillo at Chichén Itzá in Mexico is designed to cast shadows in the shape of serpents climbing the pyramid at the vernal and autumn equinoxes.

In the late Roman Empire the Sun's birthday was a holiday celebrated as Sol Invictus (literally "unconquered sun") soon after the winter solstice which may have been an antecedent to Christmas. Regarding the fixed stars, the Sun appears from Earth to revolve once a year along the ecliptic through the zodiac, and so Greek astronomers considered it to be one of the seven planets (Greek *planetes*, "wanderer"), after which the seven days of the week are named in some languages.[130] [131] [132]

Development of scientific understanding

In the early first millennium BCE, Babylonian astronomers observed that the Sun's motion along the ecliptic was not uniform, though they were unaware of why this was; it is today known that this is due to the Earth moving in an elliptic orbit around the Sun, with the Earth moving faster when it is nearer to the Sun at perihelion and moving slower when it is farther away at aphelion.[133]

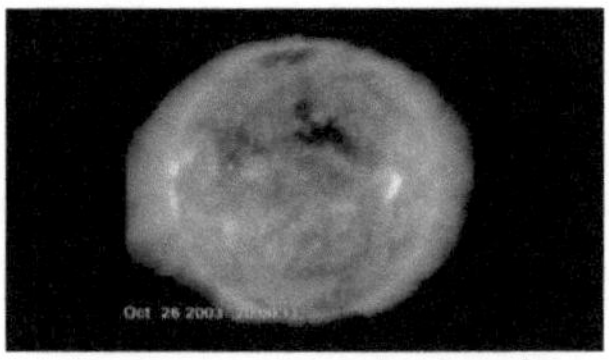

Since the discovery of sunspots by Galileo in 1609, we have continued to study the Sun.

One of the first people to offer a scientific or philosophical explanation for the Sun was the Greek philosopher Anaxagoras, who reasoned that it was a giant flaming ball of metal even larger than the Peloponnesus rather than the chariot of Helios, and that the Moon reflected the light of the Sun.[134] For teaching this heresy, he was imprisoned by the authorities and sentenced to death, though he was later released through the intervention of Pericles. Eratosthenes estimated the distance between the Earth and the Sun in the 3rd century BCE as "of stadia myriads 400 and 80000", the translation of which is ambiguous, implying either 4,080,000 stadia (755,000 km) or 804,000,000 stadia (148 to 153 million kilometers or 0.99 to 1.02 AU); the latter value is correct to within a few percent. In the 1st century CE, Ptolemy estimated the distance as 1,210 times the Earth radius, approximately 7.71 million kilometers (AU).[135]

The theory that the Sun is the center around which the planets move was first proposed by the ancient Greek Aristarchus of Samos in the 3rd century BCE, and later adopted by Seleucus of Seleucia (see Heliocentrism). This largely philosophical view was developed into fully predictive mathematical model of a heliocentric system in the 16th century by Nicolaus Copernicus. In the early 17th century, the invention of the telescope permitted detailed observations of sunspots by Thomas Harriot, Galileo Galilei and other astronomers. Galileo made some of the first known telescopic observations of sunspots and posited that they were on the surface of the Sun rather than small objects passing between the Earth and the Sun.[136] Sunspots were also observed since the Han Dynasty (206 BCE – 220 CE) by Chinese astronomers who maintained records of these observations for centuries. Averroes also provided a description of sunspots in the 12th century.[137]

Arabic astronomical contributions include Albatenius discovery that the direction of the Sun's apogee (the place in the Sun's orbit against the fixed stars where it seems to be moving slowest) is changing,.[138] (In modern heliocentric terms, this is caused by a gradual motion of the aphelion of the *Earth's* orbit). Ibn Yunus observed more than 10,000 entries for the Sun's position for many years using a large astrolabe.[139]

The transit of Venus was first observed in 1032 by Persian astronomer and polymath Avicenna, who concluded that Venus is closer to the Earth than the Sun,[140] while one of the first observations of the transit of Mercury was conducted by Ibn Bajjah in the 12th century.[141]

Sol, the Sun, from a 1550 edition of Guido Bonatti's *Liber astronomiae.*

In 1672 Giovanni Cassini and Jean Richer determined the distance to Mars and were thereby able to calculate the distance to the Sun. Isaac Newton observed the Sun's light using a prism, and showed that it was made up of light of many colors,[142] while in 1800 William Herschel discovered infrared radiation beyond the red part of the solar spectrum.[143] The 19th century saw advancement in spectroscopic studies of the Sun; Joseph von Fraunhofer recorded more than 600 absorption lines in the spectrum, the strongest of which are still often referred to as Fraunhofer lines.

In the early years of the modern scientific era, the source of the Sun's energy was a significant puzzle. Lord Kelvin suggested that the Sun was a gradually cooling liquid body that was radiating an internal store of heat.[144] Kelvin and Hermann von Helmholtz then proposed a gravitational contraction mechanism to explain the energy output. Unfortunately the resulting age estimate was only 20 million years, well short of the time span of at least 300 million years suggested by some geological discoveries of that time.[144] In 1890 Joseph Lockyer, who discovered helium in the solar spectrum, proposed a meteoritic hypothesis for the formation and evolution of the Sun.[145]

Not until 1904 was a documented solution offered. Ernest Rutherford suggested that the Sun's output could be maintained by an internal source of heat, and suggested radioactive decay as the source.[146] However, it would be Albert Einstein who would provide the essential clue to the source of the Sun's energy output with his mass-energy equivalence relation $E = mc^2$.[147]

In 1920, Sir Arthur Eddington proposed that the pressures and temperatures at the core of the Sun could produce a nuclear fusion reaction that merged hydrogen (protons) into helium nuclei, resulting in a production of energy from the net change in mass.[148] The preponderance of hydrogen in the Sun was confirmed in 1925 by Cecilia Payne using the ionization theory developed by Meghnad Saha, an Indian physicist. The theoretical concept of fusion was developed in the 1930s by the astrophysicists Subrahmanyan Chandrasekhar and Hans Bethe. Hans Bethe calculated the details of the two main energy-producing nuclear reactions that power the Sun.[149] [150]

Finally, a seminal paper was published in 1957 by Margaret Burbidge, entitled "Synthesis of the Elements in Stars".[151] The paper demonstrated convincingly that most of the elements in the universe had been synthesized by nuclear reactions inside stars, some like the Sun.

Solar space missions

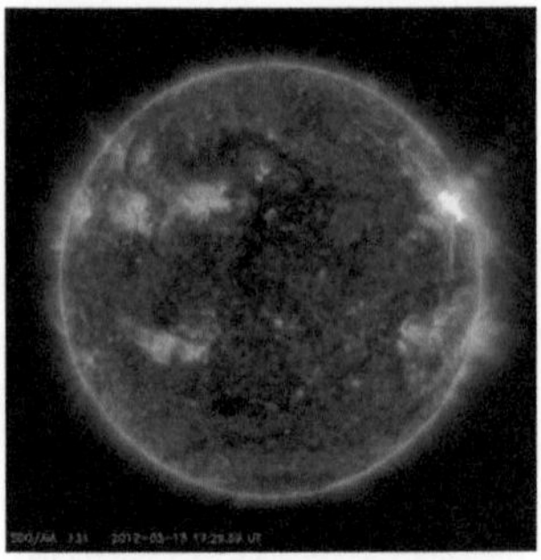

The Sun giving out a large geomagnetic storm on 1:29 pm, EST, 13th March 2012

The first satellites designed to observe the Sun were NASA's Pioneers 5, 6, 7, 8 and 9, which were launched between 1959 and 1968. These probes orbited the Sun at a distance similar to that of the Earth, and made the first detailed measurements of the solar wind and the solar magnetic field. Pioneer 9 operated for a particularly long time, transmitting data until May 1983.[152] [153]

In the 1970s, two Helios spacecraft and the Skylab Apollo Telescope Mount provided scientists with significant new data on solar wind and the solar corona. The Helios 1 and 2 probes were U.S.–German collaborations that studied the solar wind from an orbit carrying the spacecraft inside Mercury's orbit at perihelion.[154] The Skylab space station, launched by NASA in 1973, included a solar observatory module called the Apollo Telescope Mount that was operated by astronauts resident on the station.[70] Skylab made the first time-resolved observations of the solar transition region and of ultraviolet emissions from the solar corona.[70] Discoveries included the first observations of coronal mass ejections, then called "coronal transients", and of coronal holes, now known to be intimately associated with the solar wind.[154]

In 1980, the Solar Maximum Mission was launched by NASA. This spacecraft was designed to observe gamma rays, X-rays and UV radiation from solar flares during a time of high solar activity and solar luminosity. Just a few months after launch, however, an electronics failure caused the probe to go into standby mode, and it spent the next three years in this inactive state. In 1984 Space Shuttle Challenger mission STS-41C retrieved the satellite and repaired its electronics before re-releasing it into orbit. The Solar Maximum Mission subsequently acquired thousands of images of the solar corona before re-entering the Earth's atmosphere in June 1989.[155]

Launched in 1991, Japan's Yohkoh (*Sunbeam*) satellite observed solar flares at X-ray wavelengths. Mission data allowed scientists to identify several different types of flares, and demonstrated that the corona away from regions of peak activity was much more dynamic and active than had previously been supposed. Yohkoh observed an entire solar cycle but went into standby mode when an annular eclipse in 2001 caused it to lose its lock on the Sun. It was destroyed by atmospheric re-entry in 2005.[156]

One of the most important solar missions to date has been the Solar and Heliospheric Observatory, jointly built by the European Space Agency and NASA and launched on 2 December 1995.[70] Originally intended to serve a two-year mission, a mission extension through 2012 was approved in October 2009.[157] It has proven so useful that a follow-on mission, the Solar Dynamics Observatory, was launched in February 2010.[158] Situated at the Lagrangian point between the Earth and the Sun (at which the gravitational pull from both is equal), SOHO has provided a constant view of the Sun at many wavelengths since its launch.[70] Besides its direct solar observation, SOHO has enabled the discovery of a large number of comets, mostly tiny sungrazing comets which incinerate as they pass the Sun.[159]

All these satellites have observed the Sun from the plane of the ecliptic, and so have only observed its equatorial regions in detail. The Ulysses probe was launched in 1990 to study the Sun's polar regions. It first travelled to Jupiter, to "slingshot" past the planet into an orbit which would take it far above the plane of the ecliptic. Serendipitously, it was well-placed to observe the collision of Comet Shoemaker-Levy 9 with Jupiter in 1994. Once Ulysses was in its scheduled orbit, it began observing the solar wind and magnetic field strength at high solar latitudes, finding that the solar wind from high latitudes was moving at about 750 km/s which was slower than expected, and that there were large magnetic waves emerging from high latitudes which scattered galactic cosmic rays.[160]

Elemental abundances in the photosphere are well known from spectroscopic studies, but the composition of the interior of the Sun is more poorly understood. A solar wind sample return mission, Genesis, was designed to allow astronomers to directly measure the composition of solar material. Genesis returned to Earth in 2004 but was damaged by a crash landing after its parachute failed to deploy on re-entry into Earth's atmosphere. Despite severe damage, some usable samples have been recovered from the spacecraft's sample return module and are undergoing analysis.[161]

The Solar Terrestrial Relations Observatory (STEREO) mission was launched in October 2006. Two identical spacecraft were launched into orbits that cause them to (respectively) pull further ahead of and fall gradually behind the Earth. This enables stereoscopic imaging of the Sun and solar phenomena, such as coronal mass ejections.[162] [163]

The Indian Space Research Organisation has scheduled launch of a 100 kg satellite named Aditya. The satellite will be launched in 2012, and will study the dynamic Solar corona.[164]

Observation and effects

The Sun as it appears from the surface of the Earth at sunrise.

The brightness of the sun can cause pain from looking at it with the naked eye, although doing so for brief periods is not hazardous for normal, non-dilated eyes.[165] [166] Looking directly at the Sun causes phosphene visual artifacts and temporary partial blindness. It also delivers about 4 milliwatts of sunlight to the retina, slightly heating it and potentially causing damage in eyes that cannot respond properly to the brightness.[167] [168] UV exposure gradually yellows the lens of the eye over a period of years and is thought to contribute to the formation of cataracts, but this depends on general exposure to solar UV, not on whether one looks directly at the Sun.[169] Long-duration viewing of the direct Sun with the naked eye can begin to cause UV-induced, sunburn-like lesions on the retina after about 100 seconds, particularly under conditions where the UV light from the Sun is intense and well focused;[170] [171] conditions are worsened by young eyes or new lens implants (which admit more UV than aging natural eyes), Sun angles near the zenith, and observing locations at high altitude.

Viewing the Sun through light-concentrating optics such as binoculars may result in permanent damage to the retina without an appropriate filter that blocks UV and substantially dims the sunlight. Attenuating filters to view the Sun should be specifically designed for that use: some improvised filters pass UV or IR rays that can harm the eye at high brightness levels.[172] Unfiltered binoculars can deliver over 500 times as much energy to the retina as using the naked eye, killing retinal cells almost instantly. Even brief glances at the midday Sun through unfiltered binoculars can cause permanent blindness.

Partial solar eclipses are hazardous to view because the eye's pupil is not adapted to the unusually high visual contrast: the pupil dilates according to the total amount of light in the field of view, *not* by the brightest object in the field. During partial eclipses most sunlight is blocked by the Moon passing in front of the Sun, but the uncovered parts of the photosphere have the same surface brightness as during a normal day. In the overall gloom, the pupil expands from ~2 mm to ~6 mm, and each retinal cell exposed to the solar image receives about ten times more light than it would looking at the non-eclipsed Sun. This can damage or kill those cells, resulting in small permanent blind spots for the viewer.[173] The hazard is insidious for inexperienced observers and for children, because there is no perception of pain: it is not immediately obvious that one's vision is being destroyed.

During sunrise and sunset sunlight is attenuated due to Rayleigh scattering and Mie scattering from a particularly long passage through Earth's atmosphere,[174] and the Sun is sometimes faint enough to be viewed comfortably with the naked eye or safely with optics (provided there is no risk of bright sunlight suddenly appearing through a break between clouds). Hazy conditions, atmospheric dust, and high humidity contribute to this atmospheric attenuation.[175]

A rare optical phenomenon may occur shortly after sunset or before sunrise, known as a green flash. The flash is caused by light from the Sun just below the horizon being bent (usually through a temperature inversion) towards the observer. Light of shorter wavelengths (violet, blue, green) is bent more than that of longer wavelengths (yellow, orange, red) but the violet and blue light is scattered more, leaving light that is perceived as green.[176]

Ultraviolet light from the Sun has antiseptic properties and can be used to sanitize tools and water. It also causes sunburn, and has other medical effects such as the production of vitamin D. Ultraviolet light is strongly attenuated by Earth's ozone layer, so that the amount of UV varies greatly with latitude and has been partially responsible for many biological adaptations, including variations in human skin color in different regions of the globe.[177]

See also

- Advanced Composition Explorer
- Antisolar point
- List of brightest stars
- List of Solar System bodies formerly considered planets
- Solar astronomy
- Solar energy
- Sun path
- Sun dog
- Sun-Earth Day
- Sungazing

Notes

[1] Williams, D. R. (2004). "Sun Fact Sheet" (http://nssdc.gsfc.nasa.gov/planetary/factsheet/sunfact.html). NASA. . Retrieved 2010-09-27.

[2] Asplund, M.; N. Grevesse and A. J. Sauval (2006). "The new solar abundances - Part I: the observations". *Communications in Asteroseismology* **147**: 76–79. Bibcode 2006CoAst.147...76A. doi:10.1553/cia147s76.

[3] "Eclipse 99: Frequently Asked Questions" (http://education.gsfc.nasa.gov/eclipse/pages/faq.html). NASA. . Retrieved 2010-10-24.

[4] Hinshaw, G.; *et al.* (2009). "Five-year Wilkinson Microwave Anisotropy Probe observations: data processing, sky maps, and basic results". *The Astrophysical Journal Supplement Series* **180** (2): 225–245. Bibcode 2009ApJS..180..225H. doi:10.1088/0067-0049/180/2/225.

[5] Emilio, Marcelo; Kuhn, Jeff R.; Bush, Rock I.; Scholl, Isabelle F. (March 5, 2012), "Measuring the Solar Radius from Space during the 2003 and 2006 Mercury Transits" (http://arxiv.org/abs/1203.4898), *arXiv*, , retrieved March 28, 2012

[6] "Solar System Exploration: Planets: Sun: Facts & Figures" (http://web.archive.org/web/20080102034758/http://solarsystem.nasa.gov/planets/profile.cfm?Object=Sun&Display=Facts&System=Metric). NASA. Archived from the original (http://solarsystem.nasa.gov/planets/profile.cfm?Object=Sun&Display=Facts&System=Metric) on 2008-01-02. .

[7] Ko, M. (1999). "Density of the Sun" (http://hypertextbook.com/facts/1999/MayKo.shtml). In Elert, G.. *The Physics Factbook*. .

[8] "Principles of Spectroscopy" (http://www.astro.lsa.umich.edu/undergrad/Labs/spectro/short_spectro.html). University of Michigan, Astronomy Department. 30 August 2007. .

[9] Bonanno, A.; Schlattl, H.; Paternò, L. (2008). "The age of the Sun and the relativistic corrections in the EOS". *Astronomy and Astrophysics* **390** (3): 1115–1118. arXiv:astro-ph/0204331. Bibcode 2002A&A...390.1115B. doi:10.1051/0004-6361:20020749.

[10] Seidelmann, P. K.; *et al.* (2000). "Report Of The IAU/IAG Working Group On Cartographic Coordinates And Rotational Elements Of The Planets And Satellites: 2000" (http://www.hnsky.org/iau-iag.htm). . Retrieved 2006-03-22.

[11] "The Sun's Vital Statistics" (http://solar-center.stanford.edu/vitalstats.html). Stanford Solar Center. . Retrieved 2008-07-29., citing Eddy, J. (1979). *A New Sun: The Solar Results From Skylab* (http://history.nasa.gov/SP-402/contents.htm). NASA. p. 37. NASA SP-402. .

[12] "How Round is the Sun?" (http://science.nasa.gov/science-news/science-at-nasa/2008/02oct_oblatesun/). NASA. 2 October 2008. . Retrieved 7 March 2011.

[13] "First Ever STEREO Images of the Entire Sun" (http://science.nasa.gov/science-news/science-at-nasa/2011/06feb_fullsun/). NASA. 6 February 2011. . Retrieved 7 March 2011.

[14] Woolfson, M (2000). "The origin and evolution of the solar system". *Astronomy & Geophysics* **41** (1): 1.12. doi:10.1046/j.1468-4004.2000.00012.x.

[15] Basu, S.; Antia, H. M. (2008). "Helioseismology and Solar Abundances". *Physics Reports* **457** (5–6): 217. arXiv:0711.4590. Bibcode 2008PhR...457..217B. doi:10.1016/j.physrep.2007.12.002.

[16] Wilk, S. R. (2009). "The Yellow Sun Paradox" (http://www.osa-opn.org/Content/ViewFile.aspx?id=11147). *Optics & Photonics News*: 12–13. .

[17] Than, K. (2006). "Astronomers Had it Wrong: Most Stars are Single" (http://www.space.com/scienceastronomy/060130_mm_single_stars.html). Space.com. . Retrieved 2007-08-01.

[18] Lada, C. J. (2006). "Stellar multiplicity and the initial mass function: Most stars are single". *Astrophysical Journal Letters* **640** (1): L63–L66. arXiv:astro-ph/0601375. Bibcode 2006ApJ...640L..63L. doi:10.1086/503158.

[19] Burton, W. B. (1986). "Stellar parameters". *Space Science Reviews* **43** (3–4): 244–250. Bibcode 1986SSRv...43..244.. doi:10.1007/BF00190626.

[20] Bessell, M. S.; Castelli, F.; Plez, B. (1998). "Model atmospheres broad-band colors, bolometric corrections and temperature calibrations for O–M stars". *Astronomy and Astrophysics* **333**: 231–250. Bibcode 1998A&A...333..231B.

[21] "A Star with two North Poles" (http://science.nasa.gov/headlines/y2003/22apr_currentsheet.htm). *Science @ NASA*. NASA. 22 April 2003. .

[22] Riley, P.; Linker, J. A.; Mikić, Z. (2002). "Modeling the heliospheric current sheet: Solar cycle variations" (http://ulysses.jpl.nasa.gov/science/monthly_highlights/2002-July-2001JA000299.pdf). *Journal of Geophysical Research* **107** (A7): SSH 8–1. Bibcode 2002JGRA.107g.SSH8R. doi:10.1029/2001JA000299. CiteID 1136. .

[23] http://interstellar.jpl.nasa.gov/interstellar/probe/introduction/neighborhood.html, Our Local Galactic Neighborhood, NASA

[24] http://www.centauri-dreams.org/?p=14203, Into the Interstellar Void, Centauri Dreams

[25] Adams, F. C.; Graves, G.; Laughlin, G. J. M. (2004). "Red Dwarfs and the End of the Main Sequence" (http://redalyc.uaemex.mx/pdf/571/57102211.pdf). *Revista Mexicana de Astronomía y Astrofísica* **22**: 46–49. Bibcode 2004RMxAC..22...46A. .

[26] Kogut, A. et al. (1993). "Dipole Anisotropy in the COBE Differential Microwave Radiometers First-Year Sky Maps". *Astrophysical Journal* **419**: 1. arXiv:astro-ph/9312056. Bibcode 1993ApJ...419....1K. doi:10.1086/173453.

[27] "Equinoxes, Solstices, Perihelion, and Aphelion, 2000–2020" (http://aa.usno.navy.mil/data/docs/EarthSeasons.php). US Naval Observatory. 31 January 2008. . Retrieved 2009-07-17.

[28] Simon, A. (2001). *The Real Science Behind the X-Files : Microbes, meteorites, and mutants* (http://books.google.com/?id=1gXImRmz7u8C&pg=PA26&dq=bacteria+that+live+with+out+the+sun). Simon & Schuster. pp. 25–27. ISBN 0-684-85618-2. .

[29] Barnhart, Robert K. (1995) The Barnhart Concise Dictionary of Etymology, page 776. HarperCollins. ISBN 0-06-270084-7.

[30] Mallory, J. P. (1989). *In Search of the Indo-Europeans: Language, Archaeology and Myth*, page 129. Thames & Hudson. ISBN 0-500-27616-1.

[31] "Slavic_Sun_Goddess" (http://www.inhername.com/honor/names_euroeast.html/). .

[32] Barnhart, Robert K. (1995) The Barnhart Concise Dictionary of Etymology, page 778. HarperCollins. ISBN 0-06-270084-7.

[33] William Little (ed.) *Oxford Universal Dictionary*, 1955. See entry on "Sol".

[34] "Sol" (http://www.merriam-webster.com/dictionary/Sol), Merriam-Webster online, accessed July 19, 2009

[35] "Opportunity's View, Sol 959 (Vertical)" (http://www.nasa.gov/mission_pages/mer/images/pia01892.html). NASA. 2006. . Retrieved 2007-08-01.

[36] Allison, M.; Schmunk, R. (2012-08-08). "Technical Notes on Mars Solar Time as Adopted by the Mars24 Sunclock" (http://www.giss.nasa.gov/tools/mars24/help/notes.html). NASA/GISS. . Retrieved 2012-09-16.

[37] Godier, S.; Rozelot, J.-P. (2000). "The solar oblateness and its relationship with the structure of the tachocline and of the Sun's subsurface" (http://aa.springer.de/papers/0355001/2300365.pdf) (PDF). *Astronomy and Astrophysics* **355**: 365–374. Bibcode 2000A&A...355..365G.
.

[38] Jones, Geraint (16 August 2012). "Sun is the most perfect sphere ever observed in nature" (http://www.guardian.co.uk/science/2012/aug/16/sun-perfect-sphere-nature). the Guardian. . Retrieved August 19, 2012.

[39] Phillips, Kenneth J. H. (1995). *Guide to the Sun*. Cambridge University Press. pp. 78–79. ISBN 978-0-521-39788-9.

[40] Schutz, Bernard F. (2003). *Gravity from the ground up*. Cambridge University Press. pp. 98–99. ISBN 978-0-521-45506-0.

[41] In astronomical sciences, the term *heavy elements* (or *metals*) refers to all elements except hydrogen and helium.

[42] Zeilik, M.A.; Gregory, S.A. (1998). *Introductory Astronomy & Astrophysics* (4th ed.). Saunders College Publishing. p. 322. ISBN 0-03-006228-4.

[43] Falk, S. W.; Lattmer, J.M.; Margolis, S. H. (1977). "Are supernovae sources of presolar grains?". *Nature* **270** (5639): 700–701. Bibcode 1977Natur.270..700F. doi:10.1038/270700a0.

[44] Zirker, Jack B. (2002). *Journey from the Center of the Sun*. Princeton University Press. p. 11. ISBN 978-0-691-05781-1.

[45] García, R.; et al. (2007). "Tracking solar gravity modes: the dynamics of the solar core". *Science* **316** (5831): 1591–1593. Bibcode 2007Sci...316.1591G. doi:10.1126/science.1140598. PMID 17478682.

[46] Basu et al. (2009). "Fresh insights on the structure of the solar core". *Astrophysical Journal* **699** (699): 1403. arXiv:0905.0651. Bibcode 2009ApJ...699.1403B. doi:10.1088/0004-637X/699/2/1403.

[47] "NASA/Marshall Solar Physics" (http://solarscience.msfc.nasa.gov/interior.shtml). Solarscience.msfc.nasa.gov. 2007-01-18. . Retrieved 2009-07-11.

[48] Broggini, Carlo (26–28 June 2003). "Nuclear Processes at Solar Energy". *Physics in Collision*: 21. arXiv:astro-ph/0308537. Bibcode 2003phco.conf...21B.

[49] Goupil, M. J. et al. (January 2011). "Open issues in probing interiors of solar-like oscillating main sequence stars 1. From the Sun to nearly suns". *Journal of Physics: Conference Series* **271** (1): 012031. arXiv:1102.0247. Bibcode 2011JPhCS.271a2031G. doi:10.1088/1742-6596/271/1/012031

[50] p. 102, *The physical universe: an introduction to astronomy*, Frank H. Shu, University Science Books, 1982, ISBN 0-935702-05-9.

[51] Table of temperatures, power densities, luminosities by radius in the Sun (http://fusedweb.llnl.gov/CPEP/Chart_Pages/5.Plasmas/Sunlayers.html). Fusedweb.llnl.gov (1998-11-09). Retrieved on 2011-08-30.

[52] A 50 kg adult human has a volume of about 0.05 m^3, which corresponds to 13.8 watts, at the volumetric power of the solar center. This is 285 kcal/day, about 10% of the actual average caloric intake and output for humans in non-stressful conditions.

[53] Haubold, H.J.; Mathai, A.M. (May 18, 1994). "Solar Nuclear Energy Generation & The Chlorine Solar Neutrino Experiment". *Basic space science. AIP Conference Proceedings* **320**: 102. arXiv:astro-ph/9405040. Bibcode 1995AIPC..320..102H. doi:10.1063/1.47009.

[54] Myers, Steven T. (1999-02-18). "Lecture 11 – Stellar Structure I: Hydrostatic Equilibrium" (http://www.aoc.nrao.edu/~smyers/courses/astro12/L11.html). . Retrieved 15 July 2009.

[55] NASA (2007). "Ancient Sunlight" (http://sunearthday.nasa.gov/2007/locations/ttt_sunlight.php). *Technology Through Time* (50). . Retrieved 2009-06-24.

[56] Michael Stix (January 2003). "On the time scale of energy transport in the sun" (http://www.springerlink.com/content/l256u14247171u67/). *Solar Physics* **212** (1): 3–6. Bibcode 2003SoPh..212....3S. doi:10.1023/A:1022952621810. .

[57] Schlattl, H. (2001). "Three-flavor oscillation solutions for the solar neutrino problem". *Physical Review D* **64** (1): 013009. arXiv:hep-ph/0102063. Bibcode 2001PhRvD..64a3009S. doi:10.1103/PhysRevD.64.013009.

[58] "NASA – Sun" (http://mynasa.nasa.gov/worldbook/sun_worldbook.html). World Book at NASA. . Retrieved 2012-10-10.

[59] ed. by Andrew M. Soward... (2005). "The solar tachocline: Formation, stability and its role in the solar dynamo" (http://books.google.com/?id=PLNwoJ6qFoEC&pg=PA193). *Fluid dynamics and dynamos in astrophysics and geophysics reviews emerging from the Durham Symposium on Astrophysical Fluid Mechanics, July 29 to August 8, 2002*. Boca Raton: CRC Press. pp. 193–235. ISBN 978-0-8493-3355-2. .

[60] Mullan, D.J (2000). "Solar Physics: From the Deep Interior to the Hot Corona" (http://books.google.com/?id=rk5fxs55_OkC&pg=PA22). In Page, D., Hirsch, J.G.. *From the Sun to the Great Attractor*. Springer. p. 22. ISBN 978-3-540-41064-5. .

[61] Abhyankar, K.D. (1977). "A Survey of the Solar Atmospheric Models" (http://prints.iiap.res.in/handle/2248/510). *Bull. Astr. Soc. India* **5**: 40–44. Bibcode 1977BASI....5...40A. .

[62] Gibson, E.G. (1973). *The Quiet Sun*. NASA. ASIN B0006C7RS0.

[63] Shu, F.H. (1991). *The Physics of Astrophysics*. **1**. University Science Books. ISBN 0-935702-64-4.

[64] Rast, Mark; Åke Nordlund, Robert F Stein, Juri Toomre (12). "Ionization Effects in Three-Dimensional Solar Granulation Simulations" (http://adsabs.harvard.edu/full/1993ApJ...408L..53R). *The Astrophysical Journal*. . Retrieved 31 December 2012.

[65] Parnel, C.. "Discovery of Helium" (http://www-solar.mcs.st-andrews.ac.uk/~clare/Lockyer/helium.html). University of St Andrews. . Retrieved 2006-03-22.

[66] De Pontieu, B.; et al. (2007). "Chromospheric Alfvénic Waves Strong Enough to Power the Solar Wind". *Science* **318** (5856): 1574–77. Bibcode 2007Sci...318.1574D. doi:10.1126/science.1151747. PMID 18063784.

[67] Solanki, S.K.; , W. and Ayres, T. (1994). "New Light on the Heart of Darkness of the Solar Chromosphere". *Science* **263** (5143): 64–66. Bibcode 1994Sci...263...64S. doi:10.1126/science.263.5143.64. PMID 17748350.

[68] Hansteen, V.H.; Leer, E. (1997). "The role of helium in the outer solar atmosphere". *The Astrophysical Journal* **482** (1): 498–509. Bibcode 1997ApJ...482..498H. doi:10.1086/304111.

[69] Erdèlyi, R.; Ballai, I. (2007). "Heating of the solar and stellar coronae: a review". *Astron. Nachr.* **328** (8): 726–733. Bibcode 2007AN....328..726E. doi:10.1002/asna.200710803.

[70] Dwivedi, Bhola N. (2006). "Our ultraviolet Sun" (http://cs-test.ias.ac.in/cs/Downloads/article_40416.pdf) (PDF). *Current Science* **91** (5): 587–595. .

[71] Russell, C.T. (2001). "Solar wind and interplanetary magnetic filed: A tutorial" (http://www-ssc.igpp.ucla.edu/personnel/russell/papers/SolWindTutorial.pdf). In Song, Paul; Singer, Howard J. and Siscoe, George L. (PDF). *Space Weather (Geophysical Monograph)*. American Geophysical Union. pp. 73–88. ISBN 978-0-87590-984-4. .

[72] Earth's atmosphere near sea level has a particle density of about 2×10^{25} m^{-3}.

[73] A.G, Emslie; J.A., Miller (2003). "Particle Acceleration" (http://books.google.de/books?id=W_oZYFplXX0C&pg=PA275). In Dwivedi, B.N.. *Dynamic Sun*. Cambridge University Press. p. 275. ISBN 978-0-521-81057-9. .

[74] "The Distortion of the Heliosphere: Our Interstellar Magnetic Compass" (http://www.spaceref.com/news/viewpr.html?pid=16394) (Press release). European Space Agency. 2005. . Retrieved 2006-03-22.

[75] "The Mean Magnetic Field of the Sun" (http://wso.stanford.edu/#MeanField). Wilcox Solar Observatory. 2006. . Retrieved 2007-08-01.

[76] "Sci-Tech – Space – Sun flips magnetic field" (http://archives.cnn.com/2001/TECH/space/02/16/sun.flips/index.html). *CNN*. 2001-02-16. . Retrieved 2009-07-11.

[77] "The Sun Does a Flip" (http://science.nasa.gov/headlines/y2001/ast15feb_1.htm). Science.nasa.gov. 2001-02-15. . Retrieved 2009-07-11.

[78] Wang, Y.-M.; Sheeley; Sheeley, N.R. (2003). "Modeling the Sun's Large-Scale Magnetic Field during the Maunder Minimum". *The Astrophysical Journal* **591** (2): 1248–56. Bibcode 2003ApJ...591.1248W. doi:10.1086/375449.

[79] Error: Bad DOI specified!
Lodders, K. (2003). "Abundances and Condensation Temperatures of the Elements" (http://www.lpi.usra.edu/meetings/metsoc2003/pdf/

5272.pdf) (PDF). *Meteoritics & Planetary Science* **38** (suppl.): 5272. Bibcode 2003M&PSA..38.5272L. .

[80] Hansen, C.J.; Kawaler, S.A.; Trimble, V. (2004). *Stellar Interiors: Physical Principles, Structure, and Evolution* (2nd ed.). Springer. pp. 19–20. ISBN 0-387-20089-4.

[81] Hansen, C.J.; Kawaler, S.A.; Trimble, V. (2004). *Stellar Interiors: Physical Principles, Structure, and Evolution* (2nd ed.). Springer. pp. 77–78. ISBN 0-387-20089-4.

[82] Aller, L.H. (1968). "The chemical composition of the Sun and the solar system". *Proceedings of the Astronomical Society of Australia* **1**: 133. Bibcode 1968PASAu...1..133A.

[83] Hansen, C.J.; Kawaler, S.A.; Trimble, V. (2004). *Stellar Interiors: Physical Principles, Structure, and Evolution* (2nd ed.). Springer. § 9.2.3. ISBN 0-387-20089-4.

[84] Biemont, E. (1978). "Abundances of singly ionized elements of the iron group in the Sun". *Monthly Notices of the Royal Astronomical Society* **184**: 683–694. Bibcode 1978MNRAS.184..683B.

[85] Ross and Aller 1976, Withbroe 1976, Hauge and Engvold 1977, cited in Biemont 1978.

[86] Corliss and Bozman (1962 cited in Biemont 1978) and Warner (1967 cited in Biemont 1978)

[87] Smith (1976 cited in Biemont 1978)

[88] Signer and Suess 1963; Manuel 1967; Marti 1969; Kuroda and Manuel 1970; Srinivasan and Manuel 1971, all cited in Manuel and Hwaung 1983

[89] Kuroda and Manuel 1970 cited in Manuel and Hwaung 1983:7

[90] Manuel, O.K.; Hwaung, G. (1983). "Solar abundances of the elements". *Meteoritics* **18** (3): 209. Bibcode 1983Metic..18..209M.

[91] "The Largest Sunspot in Ten Years" (http://web.archive.org/web/20070823050403/http://www.gsfc.nasa.gov/gsfc/spacesci/solarexp/sunspot.htm). Goddard Space Flight Center. 30 March 2001. Archived from the original (http://www.gsfc.nasa.gov/gsfc/spacesci/solarexp/sunspot.htm) on August 23, 2007. . Retrieved 2009-07-10.

[92] "NASA Satellites Capture Start of New Solar Cycle" (http://www.physorg.com/news119271347.html). PhysOrg. 4 January 2008. . Retrieved 2009-07-10.

[93] Willson, R. C.; Hudson, H. S. (1991). "The Sun's luminosity over a complete solar cycle". *Nature* **351** (6321): 42–4. Bibcode 1991Natur.351...42W. doi:10.1038/351042a0.

[94] Lean, J.; Skumanich, A.; White, O. (1992). "Estimating the Sun's radiative output during the Maunder Minimum". *Geophysical Research Letters* **19** (15): 1591–1594. Bibcode 1992GeoRL..19.1591L. doi:10.1029/92GL01578.

[95] Mackay, R. M.; Khalil, M. A. K (2000). "Greenhouse gases and global warming" (http://books.google.com/?id=tQBS3bAX8fUC&pg=PA1&dq=solar+minimum+dendochronology). In Singh, S. N.. *Trace Gas Emissions and Plants*. Springer. pp. 1–28. ISBN 978-0-7923-6545-7. .

[96] Ehrlich, R. (2007). "Solar Resonant Diffusion Waves as a Driver of Terrestrial Climate Change". *Journal of Atmospheric and Solar-Terrestrial Physics* **69** (7): 759. arXiv:astro-ph/0701117. Bibcode 2007JASTP..69..759E. doi:10.1016/j.jastp.2007.01.005.

[97] Clark, S. (2007). "Sun's fickle heart may leave us cold" (http://environment.newscientist.com/channel/earth/mg19325884.500-suns-fickle-heart-may-leave-us-cold.html). *New Scientist* **193** (2588): 12. doi:10.1016/S0262-4079(07)60196-1. .

[98] Ribas, Ignasi (February 2010). *The Sun and stars as the primary energy input in planetary atmospheres*. "Solar and Stellar Variability: Impact on Earth and Planets, Proceedings of the International Astronomical Union, IAU Symposium". *Proceedings of the International Astronomical Union* **264**: 3–18. arXiv:0911.4872. Bibcode 2010IAUS..264....3R. doi:10.1017/S1743921309992298

[99] Amelin, Y.; Krot, A.; Hutcheon, I.; Ulyanov, A. (2002). "Lead isotopic ages of chondrules and calcium-aluminum-rich inclusions". *Science* **297** (5587): 1678–1683. Bibcode 2002Sci...297.1678A. doi:10.1126/science.1073950. PMID 12215641.

[100] Baker, J.; Bizzarro, M.; Wittig, N.; Connelly, J.; Haack, H. (2005). "Early planetesimal melting from an age of 4.5662 Gyr for differentiated meteorites". *Nature* **436** (7054): 1127–1131. Bibcode 2005Natur.436.1127B. doi:10.1038/nature03882. PMID 16121173.

[101] Error: Bad DOI specified!

[102] Goldsmith, D.; Owen, T. (2001). *The search for life in the universe* (http://books.google.com/?id=Q17NmHY6wloC&pg=PA96). University Science Books. p. 96. ISBN 978-1-891389-16-0. .

[103] Pogge, R.W. (1997). "The Once and Future Sun" (http://www.astronomy.ohio-state.edu/~pogge/Lectures/vistas97.html). *New Vistas in Astronomy* (http://www-astronomy.mps.ohio-state.edu/Vistas/). Ohio State University (Department of Astronomy). . Retrieved 2005-12-07.

[104] Sackmann, I.-J.; Boothroyd, A.I.; Kraemer, K.E. (1993). "Our Sun. III. Present and Future". *Astrophysical Journal* **418**: 457. Bibcode 1993ApJ...418..457S. doi:10.1086/173407.

[105] Schröder, K.-P.; Smith, R.C. (2008). "Distant future of the Sun and Earth revisited". *Monthly Notices of the Royal Astronomical Society* **386** (1): 155. arXiv:0801.4031. Bibcode 2008MNRAS.386..155S. doi:10.1111/j.1365-2966.2008.13022.x. See also Palmer, J. (2008). "Hope dims that Earth will survive Sun's death" (http://space.newscientist.com/article/dn13369-hope-dims-that-earth-will-survive-suns-death.html?feedId=online-news_rss20). *New Scientist*. . Retrieved 2008-03-24.

[106] Carrington, D. (2000-02-21). "Date set for desert Earth" (http://news.bbc.co.uk/1/hi/sci/tech/specials/washington_2000/649913.stm). BBC News. . Retrieved 2007-03-31.

[107] "Construction of a Composite Total Solar Irradiance (TSI) Time Series from 1978 to present" (http://www.pmodwrc.ch/pmod.php?topic=tsi/composite/SolarConstant). . Retrieved 2005-10-05.

[108] Solar radiation (http://curry.eas.gatech.edu/Courses/6140/ency/Chapter3/Ency_Atmos/Radiation_Solar.pdf)

[109] El-Sharkawi, Mohamed A. (2005). *Electric energy*. CRC Press. pp. 87–88. ISBN 978-0-8493-3078-0.

[110] "Reference Solar Spectral Irradiance: Air Mass 1.5" (http://rredc.nrel.gov/solar/spectra/am1.5/). . Retrieved 2009-11-12.

[111] Reid, M.J. (1993). "The distance to the center of the Galaxy". *Annual Review of Astronomy and Astrophysics* **31** (1): 345–372. Bibcode 1993ARA&A..31..345R. doi:10.1146/annurev.aa.31.090193.002021.

[112] Eisenhauer, F.; et al. (2003). "A Geometric Determination of the Distance to the Galactic Center". *Astrophysical Journal* **597** (2): L121–L124. arXiv:astro-ph/0306220. Bibcode 2003ApJ...597L.121E. doi:10.1086/380188.

[113] Horrobin, M.; et al. (2004). "First results from SPIFFI. I: The Galactic Center" (http://www.mpe.mpg.de/SPIFFI/preprints/first_result_an1.pdf) (PDF). *Astronomische Nachrichten* **325** (2): 120–123. Bibcode 2004AN....325...88H. doi:10.1002/asna.200310181. .

[114] Eisenhauer, F.; et al. (2005). "SINFONI in the Galactic Center: Young Stars and Infrared Flares in the Central Light-Month". *Astrophysical Journal* **628** (1): 246–259. arXiv:astro-ph/0502129. Bibcode 2005ApJ...628..246E. doi:10.1086/430667.

[115] Gehrels, Neil; Chen, Wan; Mereghetti, S. (February 25, 1993). "The Geminga supernova as a possible cause of the local interstellar bubble". *Nature* **361** (6414): 706–707. Bibcode 1993Natur.361..704B. doi:10.1038/361704a0.

[116] English, J. (2000). "Exposing the Stuff Between the Stars" (http://www.ras.ucalgary.ca/CGPS/press/aas00/pr/pr_14012000/pr_14012000map1.html) (Press release). Hubble News Desk. . Retrieved 2007-05-10.

[117] Gillman, M.; Erenler, H. (2008). "The galactic cycle of extinction". *International Journal of Astrobiology* **7** (1): 17–26. Bibcode 2008IJAsB...7...17G. doi:10.1017/S1473550408004047.

[118] Leong, S. (2002). "Period of the Sun's Orbit around the Galaxy (Cosmic Year)" (http://hypertextbook.com/facts/2002/StacyLeong.shtml). *The Physics Factbook*. . Retrieved 2007-05-10.

[119] Croswell, K. (2008). "Milky Way keeps tight grip on its neighbor" (http://space.newscientist.com/article/mg19926693.900-milky-way-keeps-tight-grip-on-its-neighbour.html). *New Scientist* (2669): 8. .

[120] Garlick, M.A. (2002). *The Story of the Solar System*. Cambridge University Press. p. 46. ISBN 0-521-80336-5.

[121] Javaraiah (2005). "Sun's retrograde motion and violation of even-odd cycle rule in sunspot activity". *Mon.Not.Roy.Astron.Soc.* **362** (4): 1311–1318. arXiv:astro-ph/0507269. Bibcode 2005MNRAS.362.1311J. doi:10.1111/j.1365-2966.2005.09403.x.

[122] Haxton, W.C. (1995). "The Solar Neutrino Problem". *Annual Review of Astronomy and Astrophysics* **33** (1): 459–504. arXiv:hep-ph/9503430. Bibcode 1995ARA&A..33..459H. doi:10.1146/annurev.aa.33.090195.002331.

[123] MacDonald, A.B. (2004). "Solar neutrinos". *New Journal of Physics* **6** (1): 121. arXiv:astro-ph/0406253. Bibcode 2004NJPh....6..121M. doi:10.1088/1367-2630/6/1/121.

[124] Ahmad, QR; et al. (2001-07-25). "Measurement of the Rate of $\nu_e + d \rightarrow p + p + e^-$ Interactions Produced by ^{8}B Solar Neutrinos at the Sudbury Neutrino Observatory". *Physical Review Letters* (American Physical Society) **87** (7): 071301. arXiv:nucl-ex/0106015. Bibcode 2001PhRvL..87g1301A. doi:10.1103/PhysRevLett.87.071301.

[125] "Sudbury Neutrino Observatory First Scientific Results" (http://www.sno.phy.queensu.ca/sno/first_results/). 2001-07-03. . Retrieved 2008-06-04.

[126] Alfvén, H. (1947). "Magneto-hydrodynamic waves, and the heating of the solar corona". *Monthly Notices of the Royal Astronomical Society* **107** (2): 211. Bibcode 1947MNRAS.107..211A.

[127] Parker, E.N. (1988). "Nanoflares and the solar X-ray corona". *Astrophysical Journal* **330** (1): 474. Bibcode 1988ApJ...330..474P. doi:10.1086/166485.

[128] Sturrock, P.A.; Uchida, Y. (1981). "Coronal heating by stochastic magnetic pumping". *Astrophysical Journal* **246** (1): 331. Bibcode 1981ApJ...246..331S. doi:10.1086/158926.

[129] Kasting, J.F.; Ackerman, T.P. (1986). "Climatic Consequences of Very High Carbon Dioxide Levels in the Earth's Early Atmosphere". *Science* **234** (4782): 1383–1385. doi:10.1126/science.11539665. PMID 11539665.

[130] "planet, n." (http://dictionary.oed.com/cgi/entry/50180718?query_type=word&queryword=planet). Oxford English Dictionary. December 2007. . Retrieved 2008-02-07. *Note: select the Etymology tab*

[131] Goldstein, Bernard R. (1997). "Saving the phenomena : the background to Ptolemy's planetary theory". *Journal for the History of Astronomy* (Cambridge (UK)) **28** (1): 1–12. Bibcode 1997JHA....28....1G.

[132] Ptolemy; Toomer, G. J. (1998). *Ptolemy's Almagest*. Princeton University Press. ISBN 978-0-691-00260-6.

[133] David Leverington (2003). *Babylon to Voyager and beyond: a history of planetary astronomy*. Cambridge University Press. pp. 6–7. ISBN 0-521-80840-5

[134] Sider, D. (1973). "Anaxagoras on the Size of the Sun". *Classical Philology* **68** (2): 128–129. doi:10.1086/365951. JSTOR 269068.

[135] Goldstein, B.R. (1967). "The Arabic Version of Ptolemy's Planetary Hypotheses". *Transactions of the American Philosophical Society* **57** (4): 9–12. doi:10.2307/1006040. JSTOR 1006040.

[136] "Galileo Galilei (1564–1642)" (http://www.bbc.co.uk/history/historic_figures/galilei_galileo.shtml). BBC. . Retrieved 2006-03-22.

[137] Ead, Hamed A.. *Averroes As A Physician*. University of Cairo.

[138] *A short History of scientific ideas to 1900*, C. Singer, Oxford University Press, 1959, p. 151.

[139] The Arabian Science, C. Ronan, pp. 201–244 in *The Cambridge Illustrated History of the World's Science*, Cambridge University Press, 1983; at pp. 213–214.

[140] Goldstein, Bernard R. (March 1972). "Theory and Observation in Medieval Astronomy". *Isis* (University of Chicago Press) **63** (1): 39–47 [44]. doi:10.1086/350839.

[141] S. M. Razaullah Ansari (2002). *History of oriental astronomy: proceedings of the joint discussion-17 at the 23rd General Assembly of the International Astronomical Union, organised by the Commission 41 (History of Astronomy), held in Kyoto, August 25–26, 1997*. Springer. p. 137. ISBN 1-4020-0657-8

[142] "Sir Isaac Newton (1643–1727)" (http://www.bbc.co.uk/history/historic_figures/newton_isaac.shtml). BBC. . Retrieved 2006-03-22.

[143] "Herschel Discovers Infrared Light" (http://coolcosmos.ipac.caltech.edu/cosmic_classroom/classroom_activities/herschel_bio.html). Cool Cosmos. . Retrieved 2006-03-22.

[144] Thomson, W. (1862). "On the Age of the Sun's Heat" (http://zapatopi.net/kelvin/papers/on_the_age_of_the_suns_heat.html). *Macmillan's Magazine* **5**: 388–393. .

[145] Lockyer, J.N. (1890). *The meteoritic hypothesis; a statement of the results of a spectroscopic inquiry into the origin of cosmical systems.* Macmillan and Co. Bibcode 1890QB981.L78......

[146] Darden, L. (1998). "The Nature of Scientific Inquiry" (http://www.philosophy.umd.edu/Faculty/LDarden/sciinq/). .

[147] Hawking, S. W. (2001). *The Universe in a Nutshell.* Bantam Books. ISBN 0-553-80202-X.

[148] "Studying the stars, testing relativity: Sir Arthur Eddington" (http://www.esa.int/esaSC/SEMDYPXO4HD_index_0.html). *Space Science.* European Space Agency. 2005. . Retrieved 2007-08-01.

[149] Bethe, H.; Critchfield, C. (1938). "On the Formation of Deuterons by Proton Combination". *Physical Review* **54** (10): 862–862. Bibcode 1938PhRv...54Q.862B. doi:10.1103/PhysRev.54.862.2.

[150] Bethe, H. (1939). "Energy Production in Stars". *Physical Review* **55** (1): 434–456. Bibcode 1939PhRv...55..434B. doi:10.1103/PhysRev.55.434.

[151] Burbidge, E.M.; Burbidge, G.R.; Fowler, W.A.; Hoyle, F. (1957). "Synthesis of the Elements in Stars". *Reviews of Modern Physics* **29** (4): 547–650. Bibcode 1957RvMP...29..547B. doi:10.1103/RevModPhys.29.547.

[152] Wade, M. (2008). "Pioneer 6-7-8-9-E" (http://www.astronautix.com/craft/pio6789e.htm). Encyclopedia Astronautica. . Retrieved 2006-03-22.

[153] "Solar System Exploration: Missions: By Target: Our Solar System: Past: Pioneer 9" (http://solarsystem.nasa.gov/missions/profile.cfm?MCode=Pioneer_09). NASA. . Retrieved 2010-10-30. "NASA maintained contact with Pioneer 9 until May 1983"

[154] Burlaga, L.F. (2001). "Magnetic Fields and plasmas in the inner heliosphere: Helios results". *Planetary and Space Science* **49** (14–15): 1619–27. Bibcode 2001P&SS...49.1619B. doi:10.1016/S0032-0633(01)00098-8.

[155] Burkepile, C. (1998). "Solar Maximum Mission Overview" (http://web.archive.org/web/20060405183758/http://web.hao.ucar.edu/public/research/svosa/smm/smm_mission.html). Archived from the original (http://web.hao.ucar.edu/public/research/svosa/smm/smm_mission.html) on April 5, 2006. . Retrieved 2006-03-22.

[156] "Result of Re-entry of the Solar X-ray Observatory "Yohkoh" (SOLAR-A) to the Earth's Atmosphere" (http://www.jaxa.jp/press/2005/09/20050913_yohkoh_e.html) (Press release). Japan Aerospace Exploration Agency. 2005. . Retrieved 2006-03-22.

[157] "Mission extensions approved for science missions" (http://sci.esa.int/science-e/www/object/index.cfm?fobjectid=45685). *ESA Science and Technology*. October 7, 2009. . Retrieved February 16, 2010.

[158] "NASA Successfully Launches a New Eye on the Sun" (http://www.nasa.gov/home/hqnews/2010/feb/HQ_10-040_SDO_launch.html). *NASA Press Release Archives.* February 11, 2010. . Retrieved February 16, 2010.

[159] "Sungrazing Comets" (http://sungrazer.nrl.navy.mil/). LASCO (US Naval Research Laboratory). . Retrieved 2009-03-19.

[160] JPL/CALTECH (2005). "Ulysses: Primary Mission Results" (http://ulysses.jpl.nasa.gov/science/mission_primary.html). NASA. . Retrieved 2006-03-22.

[161] Calaway, M.J.; Stansbery, Eileen K.; Keller, Lindsay P. (2009). "Genesis capturing the Sun: Solar wind irradiation at Lagrange 1". *Nuclear Instruments and Methods in Physics Research B* **267** (7): 1101. Bibcode 2009NIMPB.267.1101C. doi:10.1016/j.nimb.2009.01.132.

[162] "STEREO Spacecraft & Instruments" (http://www.nasa.gov/mission_pages/stereo/spacecraft/index.html). *NASA Missions.* March 8, 2006. . Retrieved May 30, 2006.

[163] Howard R. A., Moses J. D., Socker D. G., Dere K. P., Cook J. W. (2002). "Sun Earth Connection Coronal and Heliospheric Investigation (SECCHI)". *Solar Variability and Solar Physics Missions Advances in Space Research* **29** (12): 2017–2026.

[164] "The Hindu" (http://www.hindu.com/2008/01/13/stories/2008011354801000.htm). January 13, 2008. . Retrieved January 25, 2012.

[165] White, T.J.; Mainster, M.A.; Wilson, P.W.; Tips, J.H. (1971). "Chorioretinal temperature increases from solar observation". *Bulletin of Mathematical Biophysics* **33** (1): 1. doi:10.1007/BF02476660.

[166] Tso, M.O.M.; La Piana, F.G. (1975). "The Human Fovea After Sungazing". *Transactions of the American Academy of Ophthalmology and Otolaryngology* **79** (6): OP788–95. PMID 1209815.

[167] Hope-Ross, M.W.; Mahon, GJ; Gardiner, TA; Archer, DB (1993). "Ultrastructural findings in solar retinopathy". *Eye* **7** (4): 29–33. doi:10.1038/eye.1993.7. PMID 8325420.

[168] Schatz, H.; Mendelblatt, F. (1973). "Solar Retinopathy from Sun-Gazing Under Influence of LSD". *British Journal of Ophthalmology* **57** (4): 270–3. doi:10.1136/bjo.57.4.270. PMC 1214879. PMID 4707624.

[169] Chou, B.R. (2005). "Eye Safety During Solar Eclipses" (http://sunearth.gsfc.nasa.gov/eclipse/SEhelp/safety2.html). . "*While environmental exposure to UV radiation is known to contribute to the accelerated aging of the outer layers of the eye and the development of cataracts, the concern over improper viewing of the Sun during an eclipse is for the development of "eclipse blindness" or retinal burns.*"

[170] Ham, W.T. Jr.; Mueller, H.A.; Sliney, D.H. (1976). "Retinal sensitivity to damage from short wavelength light". *Nature* **260** (5547): 153. Bibcode 1976Natur.260..153H. doi:10.1038/260153a0.

[171] Ham, W.T. Jr.; Mueller, H.A.; Ruffolo, J.J. Jr.; Guerry, D. III, (1980). "Solar Retinopathy as a function of Wavelength: its Significance for Protective Eyewear". In Williams, T.P.; Baker, B.N.. *The Effects of Constant Light on Visual Processes*. Plenum Press. pp. 319–346. ISBN 0-306-40328-5.

[172] Kardos, T. (2003). *Earth science* (http://books.google.com/?id=xI6EDV_PRr4C&pg=PT102). J.W. Walch. p. 87. ISBN 978-0-8251-4500-1. .

[173] Espenak, F. (2005). "Eye Safety During Solar Eclipses" (http://sunearth.gsfc.nasa.gov/eclipse/SEhelp/safety.html). NASA. . Retrieved 2006-03-22.

[174] Haber, Jorg; Magnor, Marcus; Seidel, Hans-Peter (2005). "Physically based Simulation of Twilight Phenomena" (http://www.mpi-inf.mpg.de/~magnor/publications/tog05.pdf) (PDF). *ACM Transactions on Graphics (TOG)* **24** (4): 1353–1373. doi:10.1145/1095878.1095884. .

[175] I.G. Piggin (1972). "Diurnal asymmetries in global radiation". *Springer* **20** (1): 41–48. Bibcode 1972AMGBB..20...41P. doi:10.1007/BF02243313.

[176] "The Green Flash" (http://web.archive.org/web/20081216135504/http://www.bbc.co.uk/weather/features/understanding/greenflash.shtml). BBC. Archived from the original (http://www.bbc.co.uk/weather/features/understanding/greenflash.shtml) on 2008-12-16. . Retrieved 2008-08-10.

[177] Barsh, G.S. (2003). "What Controls Variation in Human Skin Color?". *PLoS Biology* **1** (1): e7. doi:10.1371/journal.pbio.0000027. PMC 212702. PMID 14551921.

References

Further reading

- Cohen, Richard (2010). *Chasing the Sun: the Epic Story of the Star that Gives us Life*. Simon & Schuster. ISBN 1-4000-6875-4.
- Thompson, M. J. (2004). "Solar interior: Helioseismology and the Sun's interior". *Astronomy & Geophysics* **45** (4): 21–25.

External links

- Nasa SOHO (Solar & Heliospheric Observatory) satellite (http://sohowww.nascom.nasa.gov/)
- National Solar Observatory (http://www.nso.edu/)
- Astronomy Cast: The Sun (http://www.astronomycast.com/astronomy/episode-30-the-sun-spots-and-all/)
- A collection of spectacular images of the sun from various institutions (http://www.boston.com/bigpicture/2008/10/the_sun.html) (The Boston Globe)
- Satellite observations of solar luminosity (http://www.acrim.com/)
- Sun|Trek, an educational website about the Sun (http://www.suntrek.org/)
- The Swedish 1-meter Solar Telescope, SST (http://www.solarphysics.kva.se/)
- An animated explanation of the structure of the Sun (http://alienworlds.glam.ac.uk/sunStructure.html) (University of Glamorgan)
- Animation - The Future of the Sun (http://www.youtube.com/watch?v=qpMRtvFD8ek&hl=fr)
- Solar Conveyor Belt Speeds Up (http://science.nasa.gov/headlines/y2010/12mar_conveyorbelt.htm) – NASA – images, link to report on Science

Article Sources and Contributors

(152576) 1994 NR2 *Source*: http://en.wikipedia.org/w/index.php?title=%28152576%29_1994_NR2 *Contributors*: Merovingian

Minor planet *Source*: http://en.wikipedia.org/w/index.php?title=Minor_planet *Contributors*: 3coma14, 7&6=thirteen, 786b6364, Aaron of Mpls, Aesopos, Alfons2, An Justified Wikipedian, Arsia Mons, Asikal1, Bmclaughlin9, Bryan Derksen, Chesnok, Ckatz, Colinmartin74, David Shepheard, Dicklyon, Eekster, Fartherred, Fjörgynn, Glenn L, Gpermant, GraL, Interstellar Man, JHunterJ, Jolielegal, JorisvS, Kayau, Kheider, Kwamikagami, LeaveSleaves, MER-C, Max rspct, Merovingian, Michael Hardy, Miremare, Mrbenten, Murgh, Nergaal, Niobrara, Novangelis, Nubiatech, R'n'B, RJHall, Raymondwinn, Reyk, Rjwilmsi, Rmhermen, Roentgenium111, Romit3, Rorschach, Rothorpe, Ruslik0, Serendipodous, Sfahey, Shirt58, Sobreira, Soulkeeper, Spacepotato, Spidern, Terraflorin, The Tom, The way, the truth, and the light, Thingg, Thorncrag, Tlesher, Trekphiler, Ultra two, Urhixidur, Vağinator, Velpaedia Jenkuklordanus, West.andrew.g, Wikiprodder, Wiklol, William Avery, Willking1979, Woohookitty, WooperJeff, Writtenright, Wwannsda, 80 anonymous edits

Asteroid belt *Source*: http://en.wikipedia.org/w/index.php?title=Asteroid_belt *Contributors*: 1981willy, 21655, 2T, A-giau, Abbalkea2010, Ace45954, Adam Keller, Aesopos, Ahoerstemeier, Aiken drum, Airplaneman, Akshay Lattimardi, Alansohn, AlexiusHoratius, Alfio, Alsandro, Amara, Anarchist42, Andrewlp1991, Antandrus, Apeman, Aponar Kestrel, Arsia Mons, Art LaPella, Arz1969, Avenue, Avicennasis, AxelBoldt, Baa, Bender235, Bewildebeast, Bfissa, Bigger digger, Blacx0, BlastOButter42, Bobymeltics, Bongwarrior, Bryan Derksen, Burzmali, CanadianLinuxUser, Casliber, CharlesC, Cheat2win, Chesnok, ChristTrekker, ChristopherWillis, CielProfond, CirrusHead, Cjc13, Ckatz, Closedmouth, Coalisnotcools, Colonies Chris, Crash Underride, Crazymonkey1123, DABADABADAI, Dabomb87, Danim, Davewild, Dawnseeker2000, Delrayva, Denisarona, Deon, Deor, Deuar, Discospinster, Donarreiskoffer, Dougano, DrKiernan, Dsparks000, Dspitzle, Dune010, Dwcarless, EarnestyEternity, Eccles999174, Ed Poor, Editore99, Edward130603, Egil, El C, El Cid, Emurphy42, Enviroboy, Epbr123, Everyking, Excirial, Extra999, FF2010, Falcon8765, Fcn, Feens, Felyza, Femto, Fjörgynn, Flyguy649, Folajimi, Fotaun, FrankCarroll, Frenchgizmo, Furrykef, Gaius Cornelius, Gap9551, Gareth Griffith-Jones, Garion96, Geeoharee, Gene Nygaard, Geoff Plourde, George100, Gilderien, Gilliam, Gogo Dodo, Goudzovski, Grahamec, Gtg204y, Gurch, GuyQuest, Gwillhickers, Hadseys, Hairy Dude, HalfShadow, Hamiltonstone, Harry, Hashar, HatriX, Hda3ku, Hdt83, Headbomb, HenryLi, Herostratus, Hibernian, Hiberniantears, Hike395, Hmains, Hope(N Forever), Hunnjazal, I like trains beetches, IVAN3MAN, IanOsgood, Ianare, Icairns, Ilikebakedziti, Improve, Incnis Mrsi, Iridescent, J.delanoy, JForget, JHunterJ, JR00576, Jacj, Jackfork, JamesHoadley, Jaredroberts, Jcronen1, Jeff G., Jeremy Visser, Jim1138, Jivecat, Jnestorius, JocK, Joepearson, John Reaves, John of Reading, JorisvS, Jusdafax, Jusjih, Jwh, Ke6jjj, Kejo13, Kheider, Kintetsubuffalo, Kobrabones, Kristaga, Kuru, Kwamikagami, Kyng, Lakers, Lampman, Lanthanum-138, Latulla, Lavateraguy, Lawrie88f, Leor klier, Lhopitalified, Liamdaly620, Lightmouse, Lilly Ishappy, LonelyMarble, Lotu, Luk, Luokehao, MJ94, Malinadams, Manisero399, Marcus Qwertyus, Marek69, Marksurge, Marskell, Martinwilke1980, Materialscientist, Matthewedwards, Maxamegalon2000, McGeddon, Meelar, Megalodon99, Melara..., Merovingian, Metafax1, Mgiganteus1, Michael Devore, Michael Hardy, Michaelas10, Mike s, Milonica, Moneo, Moocha, Morgan Leigh, MrRadioGuy, Mulad, Murgh, Muriel Gottrop, Murlough23, Mxn, Myanw, Myrrhlin, Narayanese, Nephelin, NerdyScienceDude, Nick C, Nick123, Nivix, NorwegianBlue, Numbers123---, Oldmanjenkins93, Oni Ookami Alfador, Ontyx, Oranjblud, Orii, Ost316, P. S. F. Freitas, P30Carl, Palica, Papa November, Paul H., Pembers, Peyre, Pharos, Philip Trueman, Piano non troppo, Pie4all88, Pinethicket, Pizza Puzzle, Pmanderson, PocketDocket, Poppy, Potatoswatter, PowerCS, Praefectorian, Qu Hongcheng, Quackers074, Quendus, Qurq, RJHall, RP88, RSStockdale, RadicalOne, Ramaksoud2000, Ran, Randomblue, Razorflame, Reaper Eternal, Res2216firestar, Retodon8, Revolución, RexNL, Reywas92, Rhowryn, Rich Farmbrough, Richhoncho, Rjwilmsi, Rmhermen, RobertG, Rogermw, Ronhjones, Rorschach, RossPatterson, Rothorpe, Rst20xx, Ruby.red.roses, Ruslik0, Rwflammang, S Whistler, SERK12, SMSshark, Sardanaphalus, Saros136, Sbig10, SchuminWeb, Scog, ScottyBerg, ScottyBoy900Q, Sdornan, Seaphoto, Sengkang, Serendipodous, Shadowcheets, Shadowjams, Shirik, Shshshsh, Siafu, Sjock, Skizzik, Skysmith, Smartech, SnappingTurtle, Snowolf, Some jerk on the Internet, Something14, Springnuts, Squirepants101, Sry85, Stephen e nelson, Subash.chandran007, Sugarfish, Sumone10154, SuperHamster, SuperSpy00bob, TRUTH316, Taed, Tbhotch, Tcncv, Tearlach, Tedickey, Template namespace initialisation script, Teog, TerraFrost, The High Fin Sperm Whale, The Rambling Man, The Tom, Thehelpfulone, Tide rolls, Tiptoety, Titus III, Tombombadil, Tpbradbury, TravisTX, Treisijs, TrevorX, Tuckerresearch, Ufviper, Uncle Dick, Urhixidur, Vanished User 0001, Viriditas, Visor, Vittsadaf, Vsmith, WRK, Waxcaptain45, Wayne Hardman, Wayne Slam, Weedwhacker128, West-J, Wetman, Whatcanbrowndo, WhiteDragon, Wiki alf, Wiki libs, Wikih101, Wikisux, Wiklol, William Avery, Wimt, Wipsenade, Wolfkeeper, WolfmanSF, Worldtraveller, Wtmitchell, X201, Xoxojoccy, Xxgurlxx, Yosef1987, Zachary, Zamphuor, Zanejune1, Zidane tribal, Zé da Silva, $_0$x, 676 anonymous edits

Asteroid *Source*: http://en.wikipedia.org/w/index.php?title=Asteroid *Contributors*: (, -- April, 10metreh, 129.128.164.xxx, 129.128.90.xxx, 15lsoucy, 1981willy, 1984, 1wonjae, 23skidoo, 28421u2232nfenfcenc, 2help, 3morand, 95j, A-giau, A4, AATTW, ABF, ACSE, AVand, Abc518, Academic Challenger, Acalamari, Access Denied, Accurizer, Adam Keller, Adambro, Addps4cat, AdjustShift, Adnergje, AgadaUrbanit, Ageekgal, Aggie747, Ahoerstemeier, Ajitem49, Ajraddatz, Alan R. Fisher, Alansohn, Ale jrb, Aleenf1, Alex.muller, Alexf, AlexiusHoratius, Alfio, Allstarecho, AllyUnion, Alsandro, Amaury, Amazins490, Amire80, Anaxial, Andonic, Andre Engels, Andrew Gray, AndrewBuck, Andy4789, AngelOfSadness, Angela, AnimAlu, AnkhMorpork, Ann Stouter, Anna512, Antandrus, Antoni Barau, Anville, Apparition11, Apple K, Arcman, Arjun01, Arkuat, Aster2, Asteroidz R not planetz, Astrobiologist, Ato, Attackpig, Avicennasis, Ayelie, Aznrules, Backslash Forwardslash, Badgernet, Banzai314, Bathanyucolkol, Beemer69, Beland, Ben-Zin, Ben.davis35, Benhocking, Bercziszani, Bevo, Bihco, BilCat, Bill37212, Billymac00, Blanchardb, Bluemask, Bob f it, Bobblewik, Bobizuruncle, Bobo192, Bogey97, Bokac1984, Bomac, Bongwarrior, Brandon, Brion VIBBER, Brockert, Bryan Derksen, Bsimmons666, CBDunkerson, CWenger, CWii, Callanecc, Callumw1, Calvin 1998, Can't sleep, clown will eat me, CapitalR, Captain Yesterday, Captain-tucker, Captainbeefart, Carinemily, Carmichael, Carson69, Casmith 789, Catapult, Catttttt2, Celarnor, Ceranthor, Charles Matthews, Charliesface, CharlotteWebb, Chase me ladies, I'm the Cavalry, Chesnok, ChiZeroOne, Chris goulet, Chuck02, Chunkylefunga, Ck lostsword, Ckatz, Ckruschke, Clovis Sangrail, Cmapm, CommonsDelinker, Conversion script, CoolbeansofGR, Coolkid1001, Courcelles, Craigboy, Creidieki, Cryptic, Css, Curps, CzarB, DARTH SIDIOUS 2, DMacks, DVD R W, DW declare, DaL33T, Daddylonglegs420, Dam sens, Danakil, Daniel, DanielCD, Danim, DanwWiki, Daranak, Dartoe, David0811, DavidLevinson, Davidiad, DazBoz, Dcoetzee, Deagle AP, Dekaels, Dendodge, Deor, Deuar, Dffgd, Diderot, Dina, Discospinster, Divebomb, Dman159753, DocWatson42, Dr. Blofeld, DrGatsby1962, Dragon of the Pants, Drbogdan, Dreish, Drift chambers, Drmies, Dstrike224, Duoduoduo, Dwadejames03, Dysepsion, E. Fokker, ERcheck, ESkog, Editor at Large, Eekster, Eggoeater, Ejosse1, El C, El Shaday, Eleassar, Electron9, Emerson7, Emijrp, Emote, Emurphy42, Enemyofthestate, EoGuy, Epbr123, Eric-Wester, Ericann726, Erik9, Etacar11, EvenGreenerFish, Evil Monkey, Evil saltine, Evvyleeds, Excirial, Eyrian, FM, Fabiform, Fabiob, Faker12345, Falcon8765, Fartherred, Femto, Ferbtastic, Ffbond, Finlay McWalter, Finngall, Firsfron, Fisherama1, Fld300b, Flex Flint, Florian Blaschke, Flyguy649, FlyingToaster, Folajimi, Foobaz, Fosnez, Fountains of Bryn Mawr, Fourohfour, Freakmighty, Freakofnurture, Fredrik, Fritzey97, Fryergreg, Furrykef, Gaius Cornelius, Giftlite, Gilliam, Glacialfox, Glen, Glenn, Globalsolidarity, Gogo Dodo, Golbez, Golgofrinchian, Gomada, GorillaWarfare, Graham87, Greenmango, GregorB, Greswik, Grstain, Gtg204y, Gtrmp, Guanaco, Gurch, Gwillhickers, H1nkles, Hackerrye, Halfway to never, Hamtechperson, Hanacy, Hanberke, Happysailor, Hapsiainen, Hdt83, Helpmet, HenryLi, Hevron1998, Hiddekel, Hiddenfromview, Hike395, Holoeconomics, HorsePunchKid, Hydrogen Iodide, Icairns, Immunize, Insanity Incarnate, Ioscius, Iridescent, Island Monkey, Ispy1981, J 1982, J appleseed2, J.delanoy, JDG, JForget, JFreeman, JHunterJ, JYolkowski, Ja 62, Jacob1207, Jadtnr1, Jake838, Jaksmata, JamieS93, Jantzen is cool, Jared Preston, JavierMC, Jiddisch, Jim.henderson, Jmrowland, Jn1611, Joefromrandb, Joelwest, John Cumbers, John Deming, John254, Jojhutton, Jon Harald Søby, Jonathunder, JorisvS, Jose77, Josh-Levin@ieee.org, Josh70597, Jpk, Jpo, Jschnur, Jusdafax, JustAddPeter, Jwrosenzweig, Jyril, Jóna Þórunn, Kazvorpal, Kborer, Keilana, Kelly Martin, Kenohki, KevWang, Kevyn, Kheider, Kier07, Kiiwiii, King of Hearts, Kingpin13, Kirill Lokshin, Kjoonlee, KleinKlio, Klilidiplomus, Kozawa, Kubigula, Kudret abi, Kurtan, Kuru, Kwamikagami, L337 kybldmstr, LALALA26, La Parka Your Car, Lamb99, Latics, Latitudinarian, Leevanjackson, Lesnail, Lexor, Lifefeed, Lightmouse, Lightyear69, Liken to a glove, LilliPuppy, Limonns, Ling.Nut, Logan, Lolihatepizza, Looxix, Loren.wilton, Loresayer, Lowe4091, Luckybeargod, Lugia2453, Luna Santin, Lupin, MATRIX, MER-C, MONGO, Macedonian, MacroDaemon, Maelnuneb, Magioladitis, Magister Mathematicae, Majorclanger, Mandarax, Mani1, Manmobile, MarcK, Marek69, Mark Foskey, Martarius, Martin451, Mas 18 dl, Masonbarge, Master Jay, Materialscientist, Mathewignash, Mattman00000, Matubo, Maury Markowitz, Mav, MavrickWeirdo, Mbc362, McSly, Mechanical digger, Mejor Los Indios, Menchi, Mendelssohn is Cool, Mephistophelian, Metalb, Metebelis, Mgmirkin, Michael Hardy, Micky kwoskie, Midnightcomm, Mike s, MikeCapone, MikeLynch, Mikenorton, Minesweeper, Minimac, MithrandirAgain, Mjuancarlo, Moomoomoo, Morwen, Moverton, Mrholybrain, Mtnerd, Muijz, Murgh, Murlough23, Mxn, Myanw, Mygerardromance, Myuphrid, NAHID, Nakon, Narcosis17, NawlinWiki, Neier, Nephelin, Nergaal, Netoholic, Newsaholic, Nick UA, Nihiltres, NiiSSA, Niobrara, Njaelkies Lea, Noclevername, Novangelis, Nubiatech, Ocaasi, Oleg Alexandrov, Olegwiki, Omicronpersei8, Orangemarlin, Oranjblud, OwenX, Oxymoron83, P. S. F. Freitas, PTJoshua, PTSE, Patanet7, Patmic2502, Patrick, Paul H., Paul Pot, Pauli133, Pedro, Perey, Pgan002, Phantomsteve, Philip Trueman, Pi@k, Pickom, Pinethicket, Piolinfax, Pit, Pizza Puzzle, PoeticVerse, Polkadotbaby0772, Polyamorph, Ponder, Porterjoh, Powered, Prashanthns, Pseudomonas, Python eggs, Q21op, Q43, Quadell, Quantpole, Quuxplusone, RFerreira, RJHall, RP88, RUL3R, RadicalOne, Rajabhattacharya, Ramesh Ramaiah, RandomCritic, Ravidreams, RazorICE, Razorflame, Rchandra, Reaper Eternal, Red Slash, RedWolf, Res2216firestar, Rettetast, RexNL, Reyk, Rich Farmbrough, Rivajunkie, Rjwilmsi, Rmhermen, RockMFR, Roentgenium111, Ron Ritzman, Rorschach, Rothorpe, Rubicon, Ruslik0, Rwflammang, Ryjaz, SJP, ST47, SU Linguist, Sagark86, Salmar, Sandhuharpreet, Sang'gre Habagat, Sapphic, Sardanaphalus, Saros136, Satori, SaveThePoint, Savemaxim, Saxsux, SchfiftyThree, Schneelocke, SchnitzelMannGreek, Sciurinæ, Sdornan, Senator Palpatine, Serendipodous, Serpentinite, Seth Ilys, Shadowjams, Shalom Yechiel, Shanes, Shirifan, Shotgun5559, Shushruth, SiliconCerebrate, SkerHawx, Skier Dude, Skizzik, SkyWalker, Slicedoranges, Slightsmile, Slon02, Snowolf, Soccer dude 07, Sodium, Sole Soul, Some jerk on the Internet, Something14, Soonhuat95, Southern Illinois SKYWARN, Sp82, Spacepotato, Spartaz, Spartenxiv, Spencer, Spitfire, Standunne, StarryGrandma, Stealingitstrongly, Stemonitis, Stephenchou0722, Steven Zhang, StitchRFC, Stormwriter, Sturmde, Stwalkerster, Subversive.sound, SummerPhD, SuperHamster, Superbeecat, Susan714, Sushi, Svetovid, Swatjester, Synchronism, Tailpig, Tassedethe, Taxman, Tbhotch, Telescopi, Template namespace initialisation script, That Guy, From That Show!, The Enlightened, The High Fin Sperm Whale, The Phoenix, The Singing Badger, The Thing That Should Not Be, TheGWO, TheKoG, Theda, Thestupidguy, ThreeBlindMice, Thue, Tide rolls, TigerShark, Tillwe, Tim123321Tim, Timir2, Timrollpickering, Timwi, Titoxd, Tjkiller555, Tom Peters, TomasBat, Tombomp, Tony Mach, Topgearstopviewer, TransUtopian, Trash boy, Tree Biting Conspiracy, Trekphiler, Tresiden, Tschild, Tyleradkins31, Ultra two, Uni223344, Untifler, Unyoyega, Uppland, Urhixidur, Vansh9988, Vary, Vağinator, Vgranucci, Vianello, Voxpuppet, Vrenator, Vsmith, Wackywace, Waltpohl, Waninge, Ward20, Wayne Hardman, Wayward, Werididiot, Werockthescienceclass, Westwood25, Wgungfu, Whitepaw, Why Not A Duck, Wiki alf, Wiki13, Wikiborg, Wikid77, Wikipelli, Wildyoda, Willhsmit, Willshark, WilyD, Wimt, Wipsenade, Wjfox2005, Wknight94, Wolfkeeper, WolfmanSF, Wwheaton, X relentless x, XJamRastafire, Xanucia, Xav71176, Xeno, Xerxes314, Xn44, Yamamoto Ichiro, Yoguys021298, Youknowandy, Yunshui, Z4ngetsu, Zidane tribal, Zsinj, Zundark, Zzzzz, ^demon, Île flottante, Саша Стефановић, Шуган, 1441 anonymous edits

Trojan (astronomy) *Source*: http://en.wikipedia.org/w/index.php?title=Trojan_%28astronomy%29 *Contributors*: AHMartin, AdnanSa, All Is One, Amakuha, AndyKali, Baseballrocks538, BatteryIncluded, BilCat, Bryan Derksen, CSZero, Christopher Cooper, Creol, Curtis Clark, Dawnseeker2000, DerBorg, Drbogdan, Enchanter, Fjörgynn, Frankie1969, Frecklefoot, GTBacchus, Goustien, Grósznyó, Hairy Dude, Hans Dunkelberg, Hellbus, Jazzlvraz, JorisvS, Kevin Myers, Kjkolb, Kwamikagami, Lavallen, Leslie Mateus, Loadmaster, Mathias-S, Maurice Carbonaro,

Medeis, Michael Hardy, Mimihitam, Mitch Ames, Njardarlogar, Ohms law, Piledhigheranddeeper, Qurq, RJHall, Rgwc, Rjwilmsi, Roentgenium111, Rothorpe, Ruslik0, Serasuna, Serendipodous, Skizzik, Spacepotato, TJRC, Tamfang, Tbhotch, Terraflorin, The Singing Badger, The Tom, TheStarmon, Theadept, Traxs7, Urhixidur, Westley Turner, Wkharrisjr, WolfmanSF, XavierGreen, Yworo, , 45 anonymous edits

Centaur (minor planet) *Source*: http://en.wikipedia.org/w/index.php?title=Centaur_%28minor_planet%29 *Contributors*: (, Acalamari, Alansohn, Argo Navis, Avinash.royyuru, Backin72, Beetstra, Bryan Derksen, CFLeon, Chesnok, Ckatz, CuriousEric, Curps, David Eppstein, Deuar, Dna-webmaster, Dread83, Duoduoduo, Egil, Eurocommuter, Filiosus's Saga, Fjörgynn, Fotaun, Gene Nygaard, Geoffrey.landis, Giftlite, Gurch, Gveret Tered, Hike395, Icairns, Ilmari Karonen, Iridia, Jmg38, John Hyams, JorisvS, Joseph Solis in Australia, Jscotti, Kbdank71, Kheider, Knowledge Seeker, Kwamikagami, Ling.Nut, Lowe4091, Maxim Razin, Mejor Los Indios, Mgmirkin, Michaelmas1957, Muma, Murgh, Njardarlogar, Novangelis, PamD, PedroPVZ, Piledhigheranddeeper, Pt, Rich Farmbrough, Rjwilmsi, Rmhermen, Roentgenium111, Ruslik0, Samdacruel, Sardanaphalus, Satori, Serendipodous, Serpentinite, Sethhater123, ShelfSkewed, Spartan, Tamfang, Template namespace initialisation script, The Singing Badger, The Tom, The shaggy one, Thebiggnome, Tom harrison, Tothebarricades.tk, Trafford09, Ufviper, Ulflarsen, Urhixidur, WolfmanSF, Δ, عباد مجاهد ديرانية, 41 anonymous edits

Kuiper belt *Source*: http://en.wikipedia.org/w/index.php?title=Kuiper_belt *Contributors*: @pple, A2-computist, Aajacksoniv, Aaron of Mpls, Abbalkea2010, Acer, Acom, AdSR, Aditya, Afroman123456789, Alfio, Altenmann, Amaurea, Andre Engels, Andres, Another Believer, Arctic.gnome, ArgGeo, Art LaPella, AshLin, Astor14, AstroHurricane001, AstroMark, AstroNomer, Ataleh, Audriusa, Av0id3r, Avsa, Barneca, Bender235, Berek, BertSen, Bfigura's puppy, Bigger digger, BlueMoonlet, Bobo192, Bolo1729, Bongwarrior, Borgx, Boston, Brandmeister, Brian0918, Brighterorange, Brim, Brion VIBBER, Bryan Derksen, Bubba73, C777, CWenger, Caerwine, Cam, Carbuncle, Card, Chesnok, Christopher Cooper, Christopher Parham, CityOfSilver, Closedmouth, Co149, Cobblet, Cometstyles, Conorobradaigh, Conversion script, Cookiehead, Craig Pemberton, Ctachme, CuriousEric, Curps, Cwolfsheep, Cyberia23, Cyp, Cyrius, DNewhall, Dabomb87, Dan Austin, Danny, Deathphoenix, Deflective, Denelson83, Desibites, Deuar, Dhum Dhum, Dna-webmaster, Downhighest, Drvancampen, Ec5618, Ed Cormany, Egil, Eippiepie33, Eleassar777, ElectricValkyrie, Ellywa, Elockard, Eob, Epbr123, Eric Wester, Ericg, Escape Orbit, Eurocommuter, Evercat, Evil Monkey, Exir Kamalabadi, FF2010, Fartherred, Fenneth, Firsfron, Fivemack, Fjörgynn, Fluorhydric, Foosher, Fredrik, GabrielEvans, Garglebutt, Geeoharee, George Ponderevo, Giftlite, Gilliam, Glenn, Gnickett1, GoingBatty, Graham87, Grtamlinb, Guessing Game, Hairy Dude, Harold f, Harp, HarryAlffa, Harryboyles, Headbomb, Heron, Hike395, Howcheng, HumphreyW, Iamunknown, Ignoranteconomist, Ilmari Karonen, Imgayftwyes, Inge-Lyubov, Iridescent, Iridia, Ishel99, J P, JDT1991, JMK, JamesFox, Jan.Kamenicek, Jann, Jaredisred, Jarry1250, JasonAQuest, Jasongetsdown, Jay Litman, Jbaugher13, Jeandré du Toit, Jeffmedkeff, Jeronimo, Jk2693, Joedeshon, Joelwest, John, John Belushi, John Hyams, John85, Jolielegal, Jon.prasad, Jonshill, Jor, JorisvS, Josh Grosse, Josiah Rowe, Jyril, KBi, Kanags, Keflavich, Kevyn, Khalid Mahmood, Kheider, Kitsunegami, Kiviuq, Kozuch, Kwamikagami, Latulla, Lestatdelc, Lightblade, Ling.Nut, Logan, LonelyMarble, LonelyPker, Looxix, Lowe4091, Luk, Lusanaherandraton, Lwalt, MK, MK8, MONGO, Madkayaker, Malhonen, Martin56, Maurice Carbonaro, Mav, Mbell, Mdotley, Mendaliv, Mentifisto, Michael Devore, Michael Hardy, Mike s, Mikegrant, Minesweeper, Modest Genius, Monz, Mooquackwooftweetmeow, Moritheil, Mrwuggs, MuZemike, Murgh, Murlough23, Myrrhlin, Mzajac, Nascarfan1,000,000,000, NeilTarrant, Nemo1986, Nemu, Nephelin, Ngebendi, Nickshanks, Nixer, Njardarlogar, Ocee, Odros, Ohconfucius, Olivier, Omnedon, OrgasGirl, Originalwana, Ottawa4ever, Outriggr, Ozone77, P. S. F. Freitas, P.T. Aufrette, Pablo-flores, Padalkar.kshitij, Palica, Pauli133, Peashy, Pharos, Philip Trueman, Philrosenberg, Piast93, Piledhigheranddeeper, Pizza Puzzle, Pmokeefe, Ponder, Poop face3432, Poppy, Princejack12, RA0808, RMHED, RP88, RadicalOne, Rasimpson, Rathanor, Rebrane, RedBLACKandBURN, RexNL, Rhodekyll, Rhort, Rich Farmbrough, Richard B, Richhoncho, Rickyrab, Rjwilmsi, Robert Merkel, RobertG, Roentgenium111, Ronhjones, RoosMargot, RossPatterson, Rothorpe, Roudy66, RoyBoy, Runningonbrains, Rursus, Ruslik0, Ruy Pugliesi, SHCarter, Sage of Ice, Sardanaphalus, Sardur, Saros136, Schneelocke, SchuminWeb, Scjessey, Scog, ScottyBoy900Q, Scruffywiki, Scwlong, Semperf, Serendipodous, Sffcorgi, Shadowjams, Shadowmorph, Shawn81, Shirulashem, Shsilver, Sich Bojan, Sideways713, Sietse Snel, Sintaku, Sir Nicholas de Mimsy-Porpington, Smartech, Some jerk on the Internet, Something14, Spellmaster, StephenBuxton, Stepheng3, Stevenj, Stirling Newberry, Sturmde, T-borg, TTE, Tamfang, Tarotcards, Tedzsee, Tegles, Template namespace initialisation script, TeunSpaans, The Other Saluton, The Thing That Should Not Be, The Tom, The Yeti, TheMadBaron, TheMightyOrb, Thegreatdr, Tide rolls, TimeCruiserMike, Timwi, Tom, TomXP411, Tomchiukc, Treisijs, Triplepickle815, Triwbe, Tronno, Tuvas, Tycho Magnetic Anomaly-1, Ufviper, Uncle G, Upquark, Urhixidur, Vahagn Petrosyan, Vanis314, Vegeta624, Vicki Rosenzweig, Vimalkalyan, Vuong Ngan Ha, Wayne Hardman, Wiki alf, Wikipelli, WilyD, WolfmanSF, Writtenonsand, Wysprgr2005, XJamRastafire, XLerate, Xiahou, Ylee, Yogirox234, Yosri, Zad68, Zero sharp, Zhatt, Саша Стефановић, باسم, ₒx, 444 anonymous edits

Trans-Neptunian object *Source*: http://en.wikipedia.org/w/index.php?title=Trans-Neptunian_object *Contributors*: -- April, 130.225.29.xxx, AKMask, Abc518, Acalamari, Acom, AdjustShift, Aeusoes1, Alec Connors, Alfio, All Is One, Anna Frodesiak, Arknascar44, BBTroll, Baric, Bender235, Berek, Bill Thayer, Brianski, Bryan Derksen, CMD Beaker, Chariot, Chesnok, Chowbok, Ckatz, ClickRick, Cohesion, Colipon, Colonies Chris, Conversion script, Craxiter, Css, Curps, Cyde, D-Notice, Daggerstab, Dante Alighieri, Delzepplin, Deuar, Dna-webmaster, Doradus, Długosz, Egil, Eleassar777, Eleland, Eog1916, Eupedia, Eurocommuter, Finlay McWalter, Firsfron, Fjörgynn, Fmarchis, Fosnez, Fournax, Freiza667, FvdP, Gjbloom, Graham87, Grammargeek, Gunter.krebs, Hairy Dude, Hans Dunkelberg, HarryAlffa, Headbomb, Hevron1998, Hike395, Horatio, HumphreyW, Iamunknown, Icairns, Ilmari Karonen, Iridia, Irpen, Ishel99, Isnow, J P, JabberWokky, Jakew, James500, Jeandré du Toit, Jkominek, Joao, Joelwest, Jonathunder, Jor, JorisvS, Josh Grosse, Jyril, Khalad, Kheider, Kungfuadam, Kwamikagami, Larry_Sanger, Lasunncty, Looxix, Luk, Marasama, Mav, Mdotley, Michael Devore, Michael Hardy, Mike Dill, Minesweeper, Mipadi, Misty Willows, Murgh, N5iln, Nbound, Nealmcb, Nephelin, Nergaal, NewEnglandYankee, Nickshanks, Night Gyr, Nixer, Ohconfucius, P. S. F. Freitas, PedroPVZ, PlasmaTime, QuantumEngineer, Qurq, RP88, RandomCritic, Reaverdrop, Renesis, Richard Arthur Norton (1958-), Richard B, Rjwilmsi, Rmhermen, Roentgenium111, Romanm, Romeu, Rothorpe, RubenGarciaHernandez, Rursus, Ruslik0, Sardanaphalus, Satori Son, Serendipodous, Sethhater123, Shanes, Siafu, Sndeep81, Some jerk on the Internet, Something14, TTE, Tachyon01, TarkusAB, Template namespace initialisation script, Tham153, The Anome, The Tom, The shaggy one, Thirty-seven, Tomaxer, UU, Uach, Ufviper, Ulric1313, Urhixidur, Vardion, Vipinhari, Vyznev Xnebara, Wassermann, Whitestarlion, Yath, Zundark, ^demon, ²¹², فرشاد اوای‌ن‌ی, ₒx, 124 anonymous edits

Sun *Source*: http://en.wikipedia.org/w/index.php?title=Sun *Contributors*: -- April, 06mkittle, 1,000,000,005k, 10metreh, 28bytes, 2D, 334a, 4twenty42o, 83d40m, 84user, 8ung3st, A. Parrot, A. di M., A3RO, A8UDI, AAG607, AJCham, AMackenzie, Abc60, Abdaal, Abeg92, Abhinavdhere, Acalamari, Ace-o-aces, Aces lead, Acet0ne, Across.The.Synapse, Acroterion, Adamlaurie, Addshore, Adoliveira, Adreamsoul, Aduda, Aeon1006, Agradman, Agunatak, Ahoerstemeier, Ahpook, Airconswitch, Airplaneman, Akajune, Akrabbim, Akriasas, Akronnick, Alansohn, Alastan, Albrozdude, AlefZet, Alessandroandcharlie, Alethiareg, Alex Bakharev, Alex.g, Alfonsobernabe, Alhutch, Alias Flood, Alinor, Allen McC., Allstarecho, Amatulic, Amcaja, AmiDaniel, Amire80, Amorymeltzer, Amyleigh47, AnOddName, Anaxial, Andigk2217, Ando228, Andrew Gray, Andrewlp1991, Andrij Kursetsky, Andy 3015, AndyZ, Andycjp, Andyroo161, Angelic Wraith, Animum, Anjouli, Ann Stouter, Anonymous Dissident, Anonymous editor, Ansumang, Antandrus, Antimatt, AolxHangover, Apemanjy, Apostrophe, Apparition11, Arakunem, Arcaani, Arctic.gnome, Arda Xi, Ardric47, ArgentTurquoise, ArglebargleIV, ArielGold, Aristotle Pagaltzis, Arjun01, Arkuat, ArmadilloFromHell, Armbrust, ArthurWeasley, Artichoker, Ashantii, Ashill, Ashmoo, Asknine, Astrobhadauria, Astronautics, Ataru, Athenean, Aude, Auslli, AviGilChaitovsky, Avia, AvicAWB, Avicennasis, Awesomechase999, Awickert, Awikiwikiwik!!!, Awolf002, AxelBoldt, Axl, AzaToth, BD2412, Babajobu, Baclough, Banes, Barbara Shack, Barek, Barron64, Barry2j, Bart133, Bartholemuel07, Barticus88, BasilRazi, Batmanand, Beaker342, Beefy SAFC, Begoon, Ben Webber, Benbest, Bender235, Benito123456789, Bennyvk, Bensaccount, Bento00, Benwildeboer, Bequinta, Berek, Bergsten, Big Bird, BigbigbigGIANTY, Bigdon128, Bijee, Bil1, Bill3025, BillMasen, BillTheBunny, Binksternet, Bkkbrad, Bkumartvm, Black Kite, Black-Velvet, Blacklans, Blake-, Blidman, Blindmonkey, Blinkismyfave, Bloodofox, Bludevilchief, Blue william, Blueblib, Bluemask, BlytheG, Bm gub, Bobblehead2, Bobblewik, Bobo192, Bogey97, Boing! said Zebedee, Bomac, Bongwarrior, Bookofjude, Boomervan123, Borislav, Bradeos Graphon, Brain40, Brandmeister, Brandmeister (old), Brandt Luke Zorn, Brendanzhang, Brent williams, BrettAllen, Brewhaha@edmc.net, Brian0918, Brian1961, Brianga, BrightStarSky, Brighterorange, Brisvegas, BritishWatcher, Brothernight, Brummer Pants, Bryan Derksen, Bucketsofg, Buddhipriya, Budhen, Bulwersator, C0nanPayne, C8sd, CERminator, CIS, CJLL Wright, CTZMSC3, CWenger, Cadby Waydell Bainbrydge, Cadiomals, Caldwell malt, Callanecc, Caltas, Caltrop, CambridgeBayWeather, Can't sleep, clown will eat me, Canadian-Bacon, CanadianCaesar, Canderson7, Cantus, Capntek, Capricorn42, Captain Wikify, CarbonRod85, Carinemily, Casliber, Caster23, Catbob625, Catgut, Causa sui, Cbennett0811, Cbrown1023, Cdc, Cdcon, CecilWard, Cedars, Celestra, CelticJobber, Cerealkiller13, Cflm001, Chan Han Xiang, Charles Matthews, CharlesC, CharlesHBennett, Chaser, Chell and the cake, Chesnok, Chibiabos, ChicXulub, Chimsta, Choppie3000, Chris the speller, ChrisWenzel, Chrisd0687, Chrislintott, Chrislk02, Christinelaura, Christopher Parham, Chriswiki, Chuchunezumi, Chun-hian, CielProfond, CioDu, Circeus, Civil Engineer III, Cj67, Ckatz, Clarityfiend, Clarknova, Clemwang, Cliff12345, ClockworkLunch, Cmapm, Coasterlover1994, CobraWiki, Cocu, Codetiger, Codex Sinaiticus, Coemgenus, CoincidentalBystander, Colinmartin74, Colonies Chris, Commander Keane, Commonbrick, CommonsDelinker, Comrade Tux, Conversion script, Coppertwig, Copytopic1, Cordless33, Cornerstone77, Corpx, Couki, Courcelles, Cp111, Cpom, Cquan, Crazycomputers, Crbnfan99, Cremepuff222, Crispmuncher, Crusadeonilliteracy, Cryptic, Crystallina, Ctifumdope, Curps, CurtisOrr, Cverlo, CyberForte, Cyde, CygnusPius, Cyktsui, Cyrus Andiron, D. Recorder, DARTH SIDIOUS 2, DO'Neil, DRTllbrg, DSachan, DV8 2XL, DVD R W, Da ninja handyman, DagnyB, Dak is back, Dalejr.3, Damërung, Dan 1024, Dancindazed, Daniel, Daniel Case, Daniel.Cardenas, DanielCD, Danny, DannyCarl, DannyZ, Dantheman531, Dar-Ape, Darktower 12345, Darrien, Darth Panda, Darthjarek, Darwinek, Dast, DaveJ7, Davewho2, David Pierce, David spector, David2032, DavidLevinson, Davidhorman, Db099221, Dbachmann, Dbfirs, Dbtfz, Dc3, Ddunau, De728631, DeWaine, DeadEyeArrow, Dean1970, Deathsquad53, Deflective, Deglr6328, Dejvid, Dekaels, Delirium, Delldot, Demmy100, DerHexer, Descendall, Desiphral, Deuar, Devein, Deviathan, Dfoofnik, Dgrant, Dhatfield, Diannaa, Diberri, Dikstr, Dillamond, Dina, DiogenesTCP, Dipankan001, Dipper3, Discospinster, Disneydude500, Dissembly, DividedByNegativeZero, Dkreisst, Dmvward, Dna-webmaster, Dogman15, Donarreiskoffer, Donbert, Dongod95, Donkeykongrider, Donmarkdixon, Dood666, Doogee, Doommaster1994, Doradus, Double sharp, DougsTech, Doulos Christos, Downthehollow, DrNortonWhite, DragonflySixtyseven, Dragons flight, Dragostanasie, Dralwik, Dratman, DreamGuy, Drift chambers, Duckbill, Duckshoot, Duke of Kent, Dureo, Dvratnam, Dvyjones, E Pluribus Anthony redux, ERcheck, Eaglizard, Ed Poor, EdgarCarpenter, Editor at Large, Eduardo Sellan III, EdwinHJ, Eeron80, Efyoo, Egil, El aprendelenguas, Elassint, Eleassar, Eleassar777, ElectronicsEnthusiast, Elizabeyth, Elkman, Elmer Clark, Elockid, Ember of Light, Emezei, Emluickyblaju, Emmett5, Emp4eva, Emperorbma, Eno-Etile, Enormousdude, Eob, Eoin5s, Epbr123, Equivocal, Ergzay, Eric Forste, Eric Kvaalen, Eric Le Bigot, Erudit 78, Escape Orbit, EscapingLife, Eschnett, Eternal Pink, Everard Proudfoot, Everyking, Everyme, Evil Monkey, EvilQueen16, Exodio, Explicit, Ezhuttukari, FF2010, FR Soliloquy, FT2, Fabricationary, Facts707, Fang Aili, Fastfission, Father Goose, Fbs. 13, FeanorStar7, Feezo, Femto, Fennec, Fenrisulfr, Ferhengvan, Fermion, Feyre, Fieari, Fifo, Fiksdal, Filam-Man, Fire 55, Fishyghost, Fizyxnrd, Fizzixman, Flag-Waving American Patriot, Flaminkight, Flammingo, Flauto Dolce, Flendersnod, Flowerpotman, Fly101, Flyguy649, Flying fish, Fonzy, Foofighter20x, Forbesb, Fotaun, Francisco Albani, Frankenpuppy, Freewayguy, FreplySpang, Friedlad, Friendly Neighbour, Frymaster, Fsdfsdfsd, Func, Funhistory, Funnybunny, Fuzheado, G Rose, G.A.S, GGT, GHe, GLaDOS, GTBacchus, Galmicmi, Gamecockfan1001, Gameseeker, Ganesh J. Acharya, Gangsters403, Gap9551, Gareth E Kegg, Garion96, Gark, Gator1, Gene Nygaard, Gene Ward Smith, Geneb1955, GeoPopID, Geoffrey.landis, GeorgeStepanek, Gerbrant, Getyourlemondaehere, Ghalas, Ghetsmith, Ghmyrtle, GianniG46, Giftlite, Gilgamesh, Gilliam, Gilo1969, Ginsengbomb, Gjd001, Gladysamuel, Glenn, Gmaxwell, Gobbleswoggler, Godlikemammal, Gogo Dodo, Goudzovski, GraemeL, Graham87, Greatparty, Green Lane, Greg L, Grokmoo, GroovySandwich, Groyolo, Gscshoyru, Guldotarpit, Gunde123456789, Gurch, GusF, Gwen Gale, Gwernol, Gzkn, Gökhan, H, H-Vergilius, Haade, Hadal, Hairy Dude, HalJor, Halsteadk, Hamiltondaniel, HandyAndy, Hankwang, Hans

Dunkelberg, HappyLogolover2011, Hapsiainen, Hardyplants, Harmil, Harold f, Harry sava, HarryAlffa, Hateme666, Haukurth, Hdt83, Headbomb, Heimstern, Hejsa, Helicoptor, Hello32020, Helvitica Bold, Henning Makholm, Henrykus, Hermione1980, Heroandgloom, Heron, Hgilbert, Hhhippo, HiEv, Hibernian, Hiberniantears, Hike395, Hive001, Hmains, Hmich176, Hoo man, HorsePunchKid, Horselover1127, Hosszuka, Howcheng, Humps, HunterX, Hunterevans, Hut 8.5, Hux, Hydrogen Iodide, I already forgot, I9o0q1, IVAN3MAN, IW.HG, Iamthedeus, IanMcGreene, IanOsgood, Iantresman, Iapetus, Iblardi, Icairns, Iceberg3k, Iced Kola, Icestorm815, Icey, IdLoveOne, Igivefakeinfo, Ignatzmice, Ihcoyc, Iikkoollpp, Ikanreed, Ilmari Karonen, Immortal Synn, Info lover, InnOmmable, InternetMeme, Interstellar Man, Invadereli94, Invincible Ninja, Ireas, Isam, Ixfd64, J.delanoy, JAF1970, JHMM13, JJB, JMK, JNW, JTN, JWB, JYolkowski, Jachin, Jacob.jose, Jagged 85, Jake Wartenberg, Jam000qaz, James Callahan, Jamesday, Jarany, Jarble, Javalizard, Jayant412, Jazdisney, Jbrezina, Jcdw14, Jdb1729, Jedidan747, Jedikaiti, Jeff G., Jeffrey O. Gustafson, Jeh, Jennavecia, Jeppesn, Jeraphine Gryphon, JeremyA, Jeroen888, Jersyko, JesseGarrett, Jhbdel, Jim Swenson, Jim77742, Jimmysauce, Jimp, JinJian, Jinxed, Jketola, Jleon, Jmcc150, Jmnbatista, Jni, Joelholdsworth, John, John D. Croft, John Mash, John Quincy Adding Machine, JohnBlackburne, JohnOwens, Johnleemk, JohnnyStJohn, Johnpaxton, Jokestress, Jon Cates, Jonik, Jonverve, JorisvS, Jose77, Joseph Solis in Australia, Josephprymak, Joshead, Joshronsen, Joshua Scott, Jossi, Joyous!, Jpastelero, Jprg1966, Jreferee, Jrockley, Jthebeatles, Juanpabl, Jujutacular, Jump off duck, Jusjih, Just James, Justice for All, Jwray, Jyril, JzG, K.lee, K8TEK, KFan II, KNM, KPH2293, Kaage, Kablammo, Karn, Karol Langner, Katalaveno, Katydidit, Kazrak, Kbk, Kbrose, Keegan, Keenan Pepper, Keilana, Keith-264, Kelvinsong, Kenj0418, Kerms, Ketiltrout, Kfc1864, Khalid Mahmood, Kharri1073, Kheider, Kier07, Kierco0619, Killdevil, Kilo-Lima, KimDabelsteinPetersen, Kimse, King Vegita, Kingturtle, Kiore, Kjetil, Klparrot, Kmac001, Knowledgeum, Koavf, KokkaShinto, Korean alpha for knowledge, Kosebamse, Kothari.sagar, Kozuch, Krash, Krawi, Krich, Krishvanth, Kristen Eriksen, Krun, Kstraka, Kukini, Kulshrax, Kungfuadam, Kungming2, Kuru, Kusunose, Kwamikagami, Kylemew, Kyonkyon, L Kensington, LOL, Laelele, Lambdoid, Landisdesign, Lapping, Last Avenue, Latitude0116, Latka, Lawstubes, Le sacre, LeaveSleaves, Lectonar, Legend78, Legolost, LeighvsOptimvsMaximvs, Lemchesvej, Lepota, Leslie Mateus, Lesonyrra, Letthetruthbetoldforalltobehold, Levineps, Lewis007, Lexicon, Lexor, Liberty Miller, Lidingo, Liftarn, Light current, Lightdarkness, Lightmouse, Lights, Ligulem, LikeLakers2, Lilac Soul, Lipis, Lir, Lmcelhiney, Loisel, LokiClock, LonelyMarble, Looxix, Lord Patrick, LordCo Centre, Louis Waweru, Lradrama, Lrdwhyt, LtNOWIS, Luk, Luna Santin, LunceFordPrefect, Lunchscale, Lunokhod, Lycurgus, Lykaestria, M1n2b3, MAN3212, MER-C, MK Dempsey, MKar, MONGO, MacGyverMagic, Maclean25, Macncc, MadGuy7023, Madhava 1947, Madjaxter, Madler, Maestlin, Magister Mathematicae, Magnus Manske, Maha ts, Maias, Maide, Majorclanger, Majorly, Makemi, Malafaya, Malcolm Farmer, Maldek, Malhonen, Mallanox, Malljaja, Man vyi, Man with two legs, Mantosh1980, Marcos, Mareino, Mark Foskey, Mark K. Jensen, MarkS, MarkSutton, Markhayward, Markjoseph125, Markus Kuhn, Markus Poessel, Marshallsumter, Marskell, Martial75, Martin451, Martins007, Martinwilke1980, Marukh18, Master Jay, Master of Puppets, MasterMind225, MasterSci, Mastorrent, Materialscientist, MatheoDJ, Matholo1, Mattd4u2nv, Matthewcieplak, Mattw998, Mattwhiteski, Maurice Carbonaro, Mav, Maximus Rex, Mayumashu, Mbariel, Mbc362, Mcorazao, Mdann52, Meaghan, Mean as custard, MearsMan, Mech Aaron, Megaguy4, Mervyn, Mesosphere, Mets501, MetsFan76, Michael C Price, Michael Devore, Michael Dinolfo, Michael Hardy, Michaelbusch, Midgrid, Mike R, Mike741, Minesweeper, Minna Sora no Shita, Miquonranger03, MisfitToys, Mithridates, Mjklin, Mjpulohanan, Mkdw, MoRsE, Modþryð, Moe Epsilon, Mojohaza1, Monado, Moncrief, Monsterblarg, Monstrabisne, Montrealais, Moop stick, Mooverb, Mopcwiki, Morganbyers, Morwen, Mourn, Moxy, Moyogo, Mr A Pinder, Mr Stephen, Mr.Chocobos, MrBell, MrFish, Mrld, Mrshaba, Mshecket, Murgh, Murlough23, Mushroom, Mutinus, Mwarren us, Mwmt1, Mxn, My76Strat, Mycroft7, Myles325a, Myrrhlin, Mysid, Mzmadmike, NERVUN, NKanngaz, NMChico24, Nabla, Nagy, Nakon, Nalumc, Natalie Erin, Natalie Hasselriis, Nathan Johnson, NatureA16, NatusRoma, NawlinWiki, Ndenison, Nehushtan, Neonumbers, Nephelin, NerdyNSK, Nergaal, Netizen, Netoholic, Netscott, NewEnglandYankee, NewProvidence, Newone, Newton2, Nicholomothy, Nick Mks, Nick Number, Nick123, Nickptar, Niczar, Nightscream, Nik42, Nikai, Ninja Wizard, Nishani 1995, Nishkid64, Nivix, Njaelkies Lea, Njál, Nlilovic, Nnemo, No Guru, Noca2plus, Noface1, Noh Boddy, Nol888, Nono64, Nopetro, Noren, Northumbrian, Novangelis, NuclearWarfare, Numbo3, Nurg, Nuttyisms, Nymbusfhs, OGRE 2, Obluis, Obradovic Goran, OfficeGirl, Ohnoitsjamie, Oldfield, Oli Filth, Oliver Manuel, Olorin28, Omicronpersei8, Oneill05 oneill, Oni Ookami Alfador, Ontopofthewall, Opelio, Opferman, Opspin, Orangutan, OrbitOne, Organics, Originalwana, Oscarthecat, Ospalh, Ossanha, Osufitchi, OverlordQ, OwenX, Oxymoron83, PAR, PSimeon, PYMontpetit, Pablothegreat85, Padajtsch-kall, Palaeovia, Parasite, Part Deux, Patrick, Patstuart, Paul Matthews, Paul venter, Paulcmnt, Pbroks13, Pciszek, Pearle, Pegasus1138, Pejmany, Pendragon5, Pepper, Peppergrower, PericlesofAthens, Peripitus, Peruvianllama, Peter Delmonte, Petri Krohn, Pgan002, Pgk, Phil Boswell, Philip Trueman, PhySusie, Piano non troppo, Piccor, PichuUmbreon, Pigduckmeatshroom, Piggysrcute123, Pigs8u3, Pinethicket, Piotrus, Pious7, Piperh, Pizza Puzzle, Pjvpjv, Pkpat2011, Planet-man828, PlanetStar, Pleroma, PoccilScript, Pocketthis, Poindexter Propellerhead, Pokathon3000, Pokefan098, Pokem0n0mekop, Pol098, Ponder, PonyToast, Poor Yorick, Postdlf, Poverty, PrincessJO, Prokillz21, PseudoSudo, Psychonaut, PurpleCoolSun, Puzl bustr, Q2op, QuackGuru, Quadell, QuartierLatin1968, Quintote, Qwertyus, Qxz, R'n'B, R1r1f2, RJAM1, RJFJR, RJHall, Rabsak, Rachelpacinda, RadicalBender, Raihan.mahmud, Rainald62, Raja99, Rakasan, Rambo forever, RandomCritic, Randomblue, Rapscallion, Raul654, Raven4x4x, Rawrthness, Raymond arritt, Razimantv, Rdsmith4, Realm of Shadows, Reaper X, RebirthThom, Redd Dragon, Redthoreau, Reign atreyu, Rejedef, Remi81992, Rerumirf, RetiredUser2, RexNL, Reyk, Reywas92, Rhombus, Riana, Rich Farmbrough, Richard L. Peterson, Richardhenwood, RickK, Rickyrab, Ricnun, Rida666, Ridwan93, Rimmo21, Ritabetta, Rjwilmsi, Rmhermen, Rnbc, Rnt20, Robert K S, Robert Merkel, RobertG, Robertburke2003, Roberto Mura, Robertvan1, Robinh, Robma, Robomaeyhem, Rocastelo, Roger M.E. Lightly, Roland Kaufmann, Roleplayer, Rominandreu, Rorschach, Rory096, Rotemliss, Rothorpe, Roxivus, RoyBoy, Rpbreen, Rschen7754, Rsheridan6, Rubber soul, Rubik-wuerfel, Rursus, Ruslik0, Rutherfordjigsaw, Rwflammang, Ryan Postlethwaite, Ryulong, SEWilco, SL93, SMC89, SQGibbon, Saaga, Sadalsuud, Safiredx, Saforrest, Sagittarian Milky Way, Saibod, Saintrain, Sakletare, Sakurambo, Sam Blacketer, Sam Clark, Sam Ellens, Samir, Samw, Samy Merchi, Sango123, Saravask, Sardanaphalus, Saros136, Satanael, Satori, SaturnCat, Savidan, Sbharris, Scarian, Scewing, SchfiftyThree, Schgooda, Schneelocke, Schutz, Schweiwikist, Schzmo, Scientizzle, Scirocco6, Scott Johnson, ScottAlanHill, Sd31415, Sdjgjhsdrhfgdfjagihdasg, Sean D Martin, SeanMack, Seazoleta, Seb-Gibbs, Secret Squirrel, Sengkang, September 11 suicide bomber, Serendipodous, SexyBern, Seyon, Sffslibrary, Shadowblackfire, Shadowlynk, Shahid hassan99, Shane1120, Shanejohnson, Shanes, ShaunES, ShaunMacPherson, Shibo77, Shijualex, Shortdude2889, Show no mercy, Showers, Shrine of Fire, Siddharthsk2000, Sidelight12, Sideways713, Sietse Snel, Simply south, Sir48, Sirex98, Siroxo, Sivakamitvm, Sivani2006, Sjakkalle, Sjb90, Skatebiker, Skeletor 0, Skizzik, Sky Attacker, Skybon, Slntprdtr, Slugger, Smartech, Smith Jones, SmokeyTheCat, Smokizzy, SnappingTurtle, Snowolf, Snoyes, Soap, Soapthgr8, Soccerisawesome, Sohmc, Solipsist, Someguy1221, Son of Somebody, Sonicology, Sophia, Soundslikealotofhooplah, South Bay, SpaceGuide, Spacepotato, Spamguy, Spangineer, SparrowsWing, SpartanVIII, Special12321, SpeedyGonsales, SpellingGuru, Spellmaster, Spencer, Speria, Spiel496, Spiff, Spike, Spinach Dip, SpookyMulder, SpuriousQ, Squallish, Squash Racket, SqueakBox, Squidgyegg, Squish7, Squizz48, Srinivas, Srleffler, StaticGull, Steel, Stefannk, StefanosKozanis, SteinbDJ, Stephan Schulz, Stephen, Stephen G. Brown, Stephenb, Sterio, Stevesysum, Stifynsemons, Stirling Newberry, Stone, Stonnman, Strait, StrangerInParadise, Stupid Corn, Sturmde, Su huynh, Subravenkat, Substar, Subsume, Subsurd, Sun Creator, Sunray, Supasheep, Supertheman, Sushant gupta, Swarm-287, Swid, Swlenz, Swotboy2000, Syncategoremata, Syrthiss, Szajci, TBadger, THEN WHO WAS PHONE?, TJDay, TRosenbaum, TUSHANT JHA, Tablizer, Tafoyad12, TaintedMustard, TallulahBelle, Talon Artaine, Tangz22, Tanthalas39, Tarquin, Tassedethe, Tasudrty, Tbhotch, TeaDrinker, TechnoFaye, Template namespace initialisation script, Terence, Terrasidius, Tetracube, TexasAndroid, That Guy, From That Show!, The Disco King, The Duke of Waltham, The Enslaver, The Firewall, The Haunted Angel, The High Fin Sperm Whale, The Letter J, The Man in Question, The Merciful, The Pelican, The Phoenix, The Rambling Man, The Snakemaster, The Sumo, The Thing That Should Not Be, The Utahraptor, The Wiki ghost, The sun is cold, TheKoG, TheOriginalSoni, TheRanger, TheScurvyEye, Thedemonhog, Theinternetsoify, Theodolite, Theophile490, Therefore, Thestick, Thisisdeansusername, ThorX13, Thricecube, Thue, Thunderboltz, Thw1309, Tiddly Tom, Tide rolls, Tifego, TigerShark, Tim R, Timb66, Timberframe, Timc, Timir Saxa, Timwi, Tins128, Tiptoety, Titoxd, Tkalayci, Tobby72, Togarida41, Tom, Tom harrison, TomasBat, Tony Sidaway, Tony1, TonySt, Toon05, Torchiest, Torf74, Torturella, Touch Of Light, Tpbradbury, Tpoore1, Trafford09, Travis Evans, Travo615, Treisijs, Tritoncis101, Triwbe, Trojancowboy, Trovatore, Trusilver, Ttbya, Turnstep, Tuxley, Twang, Tweenk, Twigge, Twilight Realm, Tycho Magnetic Anomaly-1, Tyler Lavhey, Tyrol5, Tzsch, U av probs, U-235, URMEL, USferdinand, Ufviper, Ulflund, Ulric1313, Undeadcow66, UnitedStatesian, Unknownheror, Unschool, Urhixidur, Urod, Usernamewhocares, Utcursch, Vac cutter, Vagmaster69, Vanished user 5zariu3jisj0j4irj, Vary, Vedranf, VegaDark, Velella, Velvetron, Vendetta411, VeryVerily, Vextration, Vicki Rosenzweig, Vilerage, Vinsci, Violetriga, Vrenator, Vsmith, Vt-aoe, Vystrix Nexoth, Vyznev Xnebara, WAS 4.250, WODUP, WadeSimMiser, WagByName, Waggers, Wagner, WakiMiko, Wareh, WarthogDemon, Watch37264, Wavelength, Wayne Miller, Wayward, Weeksy, Werothegreat, WesDecker, Whoiam55, Wiglaf, Wikepediamaniac, Wiki alf, WikiFan04, Wikieditor06, Wikiwee, Wikiwhat?, Wikiwikiwee, Will Aaron 6, William Allen Simpson, WilliamKF, Willking1979, Wimt, Wisq, Witbrock, Wizzard2k, Wknight94, Wombatcat, Woohookitty, Worldtraveller, WriterHound, X3nolith, XJamRastafire, Xaonon, Xaosflux, XayaneXeX, Xcindydollx, Xdef, Xerxes314, Xtraeme, Yansa, Yath, Ybbor, Yidisheryid, Yosef 52, Yousou, Ysw1987, Yuma en, Yvesnimmo, ZX81, Zahab, Zbayz, Zbxgscqf, Zelmerszoetrop, Zeneky, ZephyrAnycon, Zimbabwe, Ziza123, Zowie, Zsinj, ZuluKing, Zzuuzz, 2088 anonymous edits

Image Sources, Licenses and Contributors

Image:InnerSolarSystem-en.png *Source*: http://en.wikipedia.org/w/index.php?title=File:InnerSolarSystem-en.png *License*: unknown *Contributors*: Original uploader was Mdf at en.wikipedia

Image:Giuseppe Piazzi.jpg *Source*: http://en.wikipedia.org/w/index.php?title=File:Giuseppe_Piazzi.jpg *License*: unknown *Contributors*: Howcheng, Lalupa, Lotse, Papa November, Rentier, Rootology, Ruslik0

Image:Main belt i vs a.png *Source*: http://en.wikipedia.org/w/index.php?title=File:Main_belt_i_vs_a.png *License*: unknown *Contributors*: Piotr Deuar

Image:951 Gaspra.jpg *Source*: http://en.wikipedia.org/w/index.php?title=File:951_Gaspra.jpg *License*: unknown *Contributors*: Alaniaris, BokicaK, Papa November, Ruslik0, The Evil IP address, Tryphon, Yonidebest, 5 anonymous edits

Image:AllendeMeteorite.jpg *Source*: http://en.wikipedia.org/w/index.php?title=File:AllendeMeteorite.jpg *License*: unknown *Contributors*: User:Vesta

Image:Main belt e vs a.png *Source*: http://en.wikipedia.org/w/index.php?title=File:Main_belt_e_vs_a.png *License*: unknown *Contributors*: Original uploader was Deuar at en.wikipedia

Image:Kirkwood Gaps.svg *Source*: http://en.wikipedia.org/w/index.php?title=File:Kirkwood_Gaps.svg *License*: unknown *Contributors*: based on plot by Alan Chamberlain, JPL/Caltech

Image:zodiacal.jpg *Source*: http://en.wikipedia.org/w/index.php?title=File:Zodiacal.jpg *License*: unknown *Contributors*: Original uploader was Serendipodous at en.wikipedia

Image:Asteroid proper elements i vs e.png *Source*: http://en.wikipedia.org/w/index.php?title=File:Asteroid_proper_elements_i_vs_e.png *License*: unknown *Contributors*: Bryan Derksen, Deuar, Papa November, Ruslik0, WinstonSmith, Yann, 1 anonymous edits

Image:Dawn Flight Configuration 2.jpg *Source*: http://en.wikipedia.org/w/index.php?title=File:Dawn_Flight_Configuration_2.jpg *License*: unknown *Contributors*: GDK

Image:Asteroidsscale.jpg *Source*: http://en.wikipedia.org/w/index.php?title=File:Asteroidsscale.jpg *License*: unknown *Contributors*: NASA/JPL-Caltech/JAXA/ESA

Image:4 Vesta 1 Ceres Moon at 20 km per px.png *Source*: http://en.wikipedia.org/w/index.php?title=File:4_Vesta_1_Ceres_Moon_at_20_km_per_px.png *License*: unknown *Contributors*: NASA, composition by Urhixidur

Image:Ceres symbol.svg *Source*: http://en.wikipedia.org/w/index.php?title=File:Ceres_symbol.svg *License*: unknown *Contributors*: Aquantrum, Crei0, Galphi, Lexicon, Ramiy, Ruslik0, Sarang, Sarilho1, Urhixidur, 4 anonymous edits

Image:Ceres2.svg *Source*: http://en.wikipedia.org/w/index.php?title=File:Ceres2.svg *License*: unknown *Contributors*: Varun Kumar

Image:Ceres3.svg *Source*: http://en.wikipedia.org/w/index.php?title=File:Ceres3.svg *License*: unknown *Contributors*: Varun Kumar

Image:2Pallas symbol.svg *Source*: http://en.wikipedia.org/w/index.php?title=File:2Pallas_symbol.svg *License*: unknown *Contributors*: User:KleinKlio

Image:2 Pallas.svg *Source*: http://en.wikipedia.org/w/index.php?title=File:2_Pallas.svg *License*: unknown *Contributors*: Vectorized by

Image:Juno symbol.svg *Source*: http://en.wikipedia.org/w/index.php?title=File:Juno_symbol.svg *License*: unknown *Contributors*: Lexicon, Sarang, Urhixidur

Image:3 Juno (1).svg *Source*: http://en.wikipedia.org/w/index.php?title=File:3_Juno_(1).svg *License*: unknown *Contributors*: User:Sim0nx

Image:Symbol 3.jpg *Source*: http://en.wikipedia.org/w/index.php?title=File:Symbol_3.jpg *License*: unknown *Contributors*: Bkell, Garion96, Pickom, RP88

Image:Vesta symbol.svg *Source*: http://en.wikipedia.org/w/index.php?title=File:Vesta_symbol.svg *License*: unknown *Contributors*: G.dallorto, Lexicon, Rmhermen, Sarang, Urhixidur

Image:4 Vesta (0).svg *Source*: http://en.wikipedia.org/w/index.php?title=File:4_Vesta_(0).svg *License*: unknown *Contributors*: Roland Geider

Image:Simbolo di Vesta.svg *Source*: http://en.wikipedia.org/w/index.php?title=File:Simbolo_di_Vesta.svg *License*: unknown *Contributors*: Original uploader was Kwamikagami at en.wikipedia

Image:4 Vesta Unsimplified Symbol.svg *Source*: http://en.wikipedia.org/w/index.php?title=File:4_Vesta_Unsimplified_Symbol.svg *License*: unknown *Contributors*: Original uploader was Tooto at en.wikipedia

File:5 Astraea symbol alternate.svg *Source*: http://en.wikipedia.org/w/index.php?title=File:5_Astraea_symbol_alternate.svg *License*: unknown *Contributors*: Wilson, John

Image:5 Astraea Symbol.svg *Source*: http://en.wikipedia.org/w/index.php?title=File:5_Astraea_Symbol.svg *License*: unknown *Contributors*: User:Rmhermen

Image:6 Hebe Astronomical Symbol.svg *Source*: http://en.wikipedia.org/w/index.php?title=File:6_Hebe_Astronomical_Symbol.svg *License*: unknown *Contributors*: User:Rmhermen

Image:7 Iris Astronomical Symbol.svg *Source*: http://en.wikipedia.org/w/index.php?title=File:7_Iris_Astronomical_Symbol.svg *License*: unknown *Contributors*: Editor at Large, Ruslik0, Urhixidur, Wikiborg

Image:8 Flora Astronomical Symbol.svg *Source*: http://en.wikipedia.org/w/index.php?title=File:8_Flora_Astronomical_Symbol.svg *License*: unknown *Contributors*: Editor at Large, Urhixidur

Image:9 Metis symbol.svg *Source*: http://en.wikipedia.org/w/index.php?title=File:9_Metis_symbol.svg *License*: unknown *Contributors*: User:Bryan Derksen

File:10 Hygeia symbol alternate.svg *Source*: http://en.wikipedia.org/w/index.php?title=File:10_Hygeia_symbol_alternate.svg *License*: unknown *Contributors*: Hind, J. R.

Image:10 Hygiea Astronomical Symbol.svg *Source*: http://en.wikipedia.org/w/index.php?title=File:10_Hygiea_Astronomical_Symbol.svg *License*: unknown *Contributors*: User:Editor at Large

File:11 Parthenope symbol alternate.svg *Source*: http://en.wikipedia.org/w/index.php?title=File:11_Parthenope_symbol_alternate.svg *License*: unknown *Contributors*: Wilson, John

Image:11 Parthenope symbol.svg *Source*: http://en.wikipedia.org/w/index.php?title=File:11_Parthenope_symbol.svg *License*: unknown *Contributors*: User:Editor at Large

Image:12 Victoria symbol.svg *Source*: http://en.wikipedia.org/w/index.php?title=File:12_Victoria_symbol.svg *License*: unknown *Contributors*: User:Editor at Large

File:13 Egeria symbol.svg *Source*: http://en.wikipedia.org/w/index.php?title=File:13_Egeria_symbol.svg *License*: unknown *Contributors*: Hind, John Russell

File:14 Irene symbol.svg *Source*: http://en.wikipedia.org/w/index.php?title=File:14_Irene_symbol.svg *License*: unknown *Contributors*: John Wilson

Image:15 Eunomia symbol.svg *Source*: http://en.wikipedia.org/w/index.php?title=File:15_Eunomia_symbol.svg *License*: unknown *Contributors*: User:Bryan Derksen

File:16 Psyche symbol.svg *Source*: http://en.wikipedia.org/w/index.php?title=File:16_Psyche_symbol.svg *License*: unknown *Contributors*: J. R. Hind

Image:17 Thetis symbol.png *Source*: http://en.wikipedia.org/w/index.php?title=File:17_Thetis_symbol.png *License*: unknown *Contributors*: Symbol created by R. Luther

Image:18 Melpomene symbol.svg *Source*: http://en.wikipedia.org/w/index.php?title=File:18_Melpomene_symbol.svg *License*: unknown *Contributors*: Hind, John Russell

File:19 Fortuna symbol.svg *Source*: http://en.wikipedia.org/w/index.php?title=File:19_Fortuna_symbol.svg *License*: unknown *Contributors*: Hind, John Russell (May 12, 1823 – December 23, 1895)

Image:26 Proserpina symbol.svg *Source*: http://en.wikipedia.org/w/index.php?title=File:26_Proserpina_symbol.svg *License*: unknown *Contributors*: Symbol created by R. Luther

Image:28 Bellona symbol.svg *Source*: http://en.wikipedia.org/w/index.php?title=File:28_Bellona_symbol.svg *License*: unknown *Contributors*: User:Editor at Large

Image:29 Amphitrite symbol.svg *Source*: http://en.wikipedia.org/w/index.php?title=File:29_Amphitrite_symbol.svg *License*: unknown *Contributors*: Chauncey A. Goodrich and Noah Porter, editors

Image:35 Leukothea symbol.png *Source*: http://en.wikipedia.org/w/index.php?title=File:35_Leukothea_symbol.png *License*: unknown *Contributors*: G. Rumker and C.A.F. Peters

Image:37 Fides symbol.svg *Source*: http://en.wikipedia.org/w/index.php?title=File:37_Fides_symbol.svg *License*: unknown *Contributors*: R. Luther

Image:243 ida.jpg *Source*: http://en.wikipedia.org/w/index.php?title=File:243_ida.jpg *License*: unknown *Contributors*: NASA/JPL

Image:Asteroid 2004 FH.gif *Source*: http://en.wikipedia.org/w/index.php?title=File:Asteroid_2004_FH.gif *License*: unknown *Contributors*: Images obtained by Stefano Sposetti, Switzerland on March 18, 2004. Animation made Raoul Behrend, Geneva Observatory, Switzerland.

Image:Moon and Asteroids 1 to 10.svg *Source*: http://en.wikipedia.org/w/index.php?title=File:Moon_and_Asteroids_1_to_10.svg *License*: unknown *Contributors*: User:Vystrix Nexoth

Image:Ceres optimized.jpg *Source*: http://en.wikipedia.org/w/index.php?title=File:Ceres_optimized.jpg *License*: unknown *Contributors*: NASA, ESA, J. Parker (Southwest Research Institute), P. Thomas (Cornell University), and L. McFadden (University of Maryland, College Park)

Image:masses of asteroids vs main belt.png *Source*: http://en.wikipedia.org/w/index.php?title=File:Masses_of_asteroids_vs_main_belt.png *License*: unknown *Contributors*: User:Kwamikagami

Image:(253) mathilde.jpg *Source*: http://en.wikipedia.org/w/index.php?title=File:(253)_mathilde.jpg *License*: unknown *Contributors*: NASA

Image:433eros.jpg *Source*: http://en.wikipedia.org/w/index.php?title=File:433eros.jpg *License*: unknown *Contributors*: NASA/NEAR Project (JHU/APL).

File:Dawn-image-070911.jpg *Source*: http://en.wikipedia.org/w/index.php?title=File:Dawn-image-070911.jpg *License*: unknown *Contributors*: NASA/JPL-Caltech/UCLA/MPS/DLR/IDA

File:Lagrange very massive.svg *Source*: http://en.wikipedia.org/w/index.php?title=File:Lagrange_very_massive.svg *License*: unknown *Contributors*: Antonsusi, EnEdC, JorisvS, Pieter Kuiper, Tano4595, 2 anonymous edits

File:InnerSolarSystem-en.png *Source*: http://en.wikipedia.org/w/index.php?title=File:InnerSolarSystem-en.png *License*: unknown *Contributors*: Original uploader was Mdf at en.wikipedia

Image:Outersolarsystem objectpositions labels comp.png *Source*: http://en.wikipedia.org/w/index.php?title=File:Outersolarsystem_objectpositions_labels_comp.png *License*: unknown *Contributors*: 84user, Kaldari, Peteforsyth, Poppy, Venkat.athma, Wikibob, WilyD, مهربان, 7 anonymous edits

Image:TheKuiperBelt 42AU Centaurs.svg *Source*: http://en.wikipedia.org/w/index.php?title=File:TheKuiperBelt_42AU_Centaurs.svg *License*: unknown *Contributors*: User:Eurocommuter

File:AsbolA.GIF *Source*: http://en.wikipedia.org/w/index.php?title=File:AsbolA.GIF *License*: unknown *Contributors*: Aldo Vitagliano

Image:TheKuiperBelt Albedo and Color.svg *Source*: http://en.wikipedia.org/w/index.php?title=File:TheKuiperBelt_Albedo_and_Color.svg *License*: unknown *Contributors*: User:Eurocommuter

Image:Comet38P2067.gif *Source*: http://en.wikipedia.org/w/index.php?title=File:Comet38P2067.gif *License*: unknown *Contributors*: Orbit Viewer applet originally written and kindly provided by Osamu Ajiki (AstroArts), and further modified by Ron Baalke (JPL). Tweaks by Kevin Heider.

File:Outersolarsystem objectpositions labels comp.png *Source*: http://en.wikipedia.org/w/index.php?title=File:Outersolarsystem_objectpositions_labels_comp.png *License*: unknown *Contributors*: 84user, Kaldari, Peteforsyth, Poppy, Venkat.athma, Wikibob, WilyD, مهربان, 7 anonymous edits

File:GerardKuiper.jpg *Source*: http://en.wikipedia.org/w/index.php?title=File:GerardKuiper.jpg *License*: unknown *Contributors*: Dodo, Miraceti, Ruslik0, Svdmolen

File:Maunatele.jpg *Source*: http://en.wikipedia.org/w/index.php?title=File:Maunatele.jpg *License*: unknown *Contributors*: NASA

File:Lhborbits.png *Source*: http://en.wikipedia.org/w/index.php?title=File:Lhborbits.png *License*: unknown *Contributors*: Jan.Kamenicek, JorisvS, MichaelSchoenitzer, Poppy, Rathorius, Stanlekub, 3 anonymous edits

File:Dust Models Paint Alien's View of Solar System.ogv *Source*: http://en.wikipedia.org/w/index.php?title=File:Dust_Models_Paint_Alien's_View_of_Solar_System.ogv *License*: unknown *Contributors*: NASA

File:TheKuiperBelt 75AU All.svg *Source*: http://en.wikipedia.org/w/index.php?title=File:TheKuiperBelt_75AU_All.svg *License*: unknown *Contributors*: User:Eurocommuter

File:TheKuiperBelt classes-en.svg *Source*: http://en.wikipedia.org/w/index.php?title=File:TheKuiperBelt_classes-en.svg *License*: unknown *Contributors*: User:Eurocommuter, User:Lilyu

File:ExampleUpdatedHistogramOfTNOsemimajoraxii.png *Source*: http://en.wikipedia.org/w/index.php?title=File:ExampleUpdatedHistogramOfTNOsemimajoraxii.png *License*: unknown *Contributors*: User:WilyD

File:2003 UB313 near-infrared spectrum.gif *Source*: http://en.wikipedia.org/w/index.php?title=File:2003_UB313_near-infrared_spectrum.gif *License*: unknown *Contributors*: Admrboltz, Adoniscik, Bryan Derksen, Chromosome, EDUCA33E, Ewen, Jan.Kamenicek, Orionist, Poppy, Poulpy, مهربان, 1 anonymous edits

File:TheKuiperBelt PowerLaw2.svg *Source*: http://en.wikipedia.org/w/index.php?title=File:TheKuiperBelt_PowerLaw2.svg *License*: unknown *Contributors*: User:Eurocommuter

File:TheKuiperBelt Projections 100AU Classical SDO.svg *Source*: http://en.wikipedia.org/w/index.php?title=File:TheKuiperBelt_Projections_100AU_Classical_SDO.svg *License*: unknown *Contributors*: User:Eurocommuter

File:Triton moon mosaic Voyager 2 (large).jpg *Source*: http://en.wikipedia.org/w/index.php?title=File:Triton_moon_mosaic_Voyager_2_(large).jpg *License*: unknown *Contributors*: NASA / Jet Propulsion Lab / U.S. Geological Survey

File:EightTNOs.png *Source*: http://en.wikipedia.org/w/index.php?title=File:EightTNOs.png *License*: unknown *Contributors*: User:Lexicon

File:New horizons Pluto.jpg *Source*: http://en.wikipedia.org/w/index.php?title=File:New_horizons_Pluto.jpg *License*: unknown *Contributors*: Bryan Derksen, Dream out loud, Kukanotas, Ruslik0, SchuminWeb, ShakataGaNai, 2 anonymous edits

File:Kuiper belt remote.jpg *Source*: http://en.wikipedia.org/w/index.php?title=File:Kuiper_belt_remote.jpg *License*: unknown *Contributors*: User:Audriusa

Image:TheTransneptunians 73AU.svg *Source*: http://en.wikipedia.org/w/index.php?title=File:TheTransneptunians_73AU.svg *License*: unknown *Contributors*: User:Eurocommuter

Image:TheTransneptunians Color Distribution.svg *Source*: http://en.wikipedia.org/w/index.php?title=File:TheTransneptunians_Color_Distribution.svg *License*: unknown *Contributors*: User:Eurocommuter

Image:Mesoplanets.svg *Source*: http://en.wikipedia.org/w/index.php?title=File:Mesoplanets.svg *License*: unknown *Contributors*: Original uploader was Lasunncty at en.wikipedia

Image:TheTransneptunians Size Albedo Color.svg *Source*: http://en.wikipedia.org/w/index.php?title=File:TheTransneptunians_Size_Albedo_Color.svg *License*: unknown *Contributors*: Chesnok, Eurocommuter, Josex.es, Ldanielb00, Poulpy, 9 anonymous edits

Image:Sun symbol.svg *Source*: http://en.wikipedia.org/w/index.php?title=File:Sun_symbol.svg *License*: unknown *Contributors*: Andrea Coppola, AnonMoos, Aquantrum, Artem Karimov, Bobcats, Bvs-aca, Er Komandante, Fibonacci, Juiced lemon, Liftarn, Luccas, Melian, Nagy, Roomba, Rursus, Samuel Grant, Sarang, Sergey Liverko, WolfgangRieger, Wst, 16 anonymous edits

File:The_Sun_by_the_Atmospheric_Imaging_Assembly_of_NASA's_Solar_Dynamics_Observatory_-_20100819.jpg *Source*: http://en.wikipedia.org/w/index.php?title=File:The_Sun_by_the_Atmospheric_Imaging_Assembly_of_NASA's_Solar_Dynamics_Observatory_-_20100819.jpg *License*: unknown *Contributors*: NASA/SDO (AIA).

File:Incandescent Sun.ogv *Source*: http://en.wikipedia.org/w/index.php?title=File:Incandescent_Sun.ogv *License*: unknown *Contributors*: NASA/Goddard Space Flight Center

File:Sun poster.svg *Source*: http://en.wikipedia.org/w/index.php?title=File:Sun_poster.svg *License*: unknown *Contributors*: User:Kelvinsong

File:Sun parts big.jpg *Source*: http://en.wikipedia.org/w/index.php?title=File:Sun_parts_big.jpg *License*: unknown *Contributors*: Project leader: Dr. Jim Lochner; Curator: Meredith Gibb; Responsible NASA Official:Phil Newman

File:EffectiveTemperature 300dpi e.png *Source*: http://en.wikipedia.org/w/index.php?title=File:EffectiveTemperature_300dpi_e.png *License*: unknown *Contributors*: user:Sch

File:Solar eclipse 1999 4 NR.jpg *Source*: http://en.wikipedia.org/w/index.php?title=File:Solar_eclipse_1999_4_NR.jpg *License*: unknown *Contributors*: user:Lviatour

File:171879main LimbFlareJan12 lg.jpg *Source*: http://en.wikipedia.org/w/index.php?title=File:171879main_LimbFlareJan12_lg.jpg *License*: unknown *Contributors*: Hinode JAXA/NASA

File:Sun - August 1, 2010.jpg *Source*: http://en.wikipedia.org/w/index.php?title=File:Sun_-_August_1,_2010.jpg *License*: unknown *Contributors*: NASA/SDO/AIA

File:Heliospheric-current-sheet.gif *Source*: http://en.wikipedia.org/w/index.php?title=File:Heliospheric-current-sheet.gif *License*: unknown *Contributors*: Werner Heil (see "other version" below).

File:Sun STEREO 4dec2006 lrg.jpg *Source*: http://en.wikipedia.org/w/index.php?title=File:Sun_STEREO_4dec2006_lrg.jpg *License*: unknown *Contributors*: NASA (Original uploader was Tainter at en.wikipedia)

File:Solar-cycle-data.png *Source*: http://en.wikipedia.org/w/index.php?title=File:Solar-cycle-data.png *License*: unknown *Contributors*: Beland, Dragons flight, F1jmm, Lampman, Lissajous, Mgc8, Nils Simon, WikipediaMaster, Xenoforme, Xiong Chiamiov, 4 anonymous edits

File:Sunspot-number.png *Source*: http://en.wikipedia.org/w/index.php?title=File:Sunspot-number.png *License*: unknown *Contributors*: Leland McInnes at en.wikipedia

File:Solar evolution (English).svg *Source*: http://en.wikipedia.org/w/index.php?title=File:Solar_evolution_(English).svg *License*: unknown *Contributors*: User:RJHall

File:Solar Life Cycle.svg *Source*: http://en.wikipedia.org/w/index.php?title=File:Solar_Life_Cycle.svg *License*: unknown *Contributors*: User:Oliverbeatson

File:Comparison sun seen from planets.svg *Source*: http://en.wikipedia.org/w/index.php?title=File:Comparison_sun_seen_from_planets.svg *License*: unknown *Contributors*: User:Nevermore4ever

File:Actual Sunset.jpg *Source*: http://en.wikipedia.org/w/index.php?title=File:Actual_Sunset.jpg *License*: unknown *Contributors*: Pocketthis

File:Solar system barycenter.svg *Source*: http://en.wikipedia.org/w/index.php?title=File:Solar_system_barycenter.svg *License*: unknown *Contributors*: User:Rubik-wuerfel

File:236084main MilkyWay-full-annotated.jpg *Source*: http://en.wikipedia.org/w/index.php?title=File:236084main_MilkyWay-full-annotated.jpg *License*: unknown *Contributors*: NASA/JPL-Caltech/R. Hurt

File:Solvogn.jpg *Source*: http://en.wikipedia.org/w/index.php?title=File:Solvogn.jpg *License*: unknown *Contributors*: Glenn, Kairios, Kersti Nebelsiek, Malene, Winterkind

File:Observing The Sun.OGG *Source*: http://en.wikipedia.org/w/index.php?title=File:Observing_The_Sun.OGG *License*: unknown *Contributors*: NASA

File:Sun-bonatti.png *Source*: http://en.wikipedia.org/w/index.php?title=File:Sun-bonatti.png *License*: unknown *Contributors*: Smerdis of Tlön

File:Sunspots and Solar Flares.jpg *Source*: http://en.wikipedia.org/w/index.php?title=File:Sunspots_and_Solar_Flares.jpg *License*: unknown *Contributors*: Dipankan001

File:Actual Sunrise.jpeg *Source*: http://en.wikipedia.org/w/index.php?title=File:Actual_Sunrise.jpeg *License*: unknown *Contributors*: ComputerHotline, Pocketthis

MIX
Papier aus verantwortungsvollen Quellen
Paper from responsible sources
FSC® C105338

FSC
www.fsc.org

Printed by Books on Demand GmbH, Norderstedt / Germany